JONATHAN OGDEN ARMOUR

THE PACKERS
THE PRIVATE CAR LINES
AND THE PEOPLE

Elibron Classics
www.elibron.com

Elibron Classics series.

ISBN 1-4021-6010-0 (paperback)
ISBN 1-4212-7897-9 (hardcover)

This Elibron Classics Replica Edition is an unabridged facsimile
of the edition published in 1906 by Henry Altemus Company, Philadelphia.

THE PACKERS
THE PRIVATE-CAR LINES
AND THE PEOPLE

PHILIP D. ARMOUR

THE PACKERS
THE PRIVATE CAR LINES
AND THE PEOPLE

BY

J. OGDEN ARMOUR

ILLUSTRATED

PHILADELPHIA
HENRY ALTEMUS COMPANY

Published, June, 1906

FOREWORD

THE writing of the articles which have appeared in the *Saturday Evening Post*, and which forms the basis of this book, was undertaken with great reluctance. First, because of an entire lack of any disposition to thrust my views regarding any subject upon the public; second, because I was equally undisposed to assume to speak for anybody but myself. On the other hand, I could not deny the existence in the public mind of a feeling that the traditional corporation policy of silence under attack is sometimes a tacit confession to the truthfulness of the charges brought about by hostile and mistaken agitators.

Finding myself the responsible head of a business founded by a father who had put into its upbuilding the best energies of a long, active life and a very considerable genius for affairs, I found it increasingly difficult to keep silent under a long series of onslaughts by the professional agitators of the country, knowing those attacks to be unfair, unjust, untrue, and in most cases maliciously bitter. When at last this condition forced upon me the conviction that I must depart from all traditions and customs of corporation practice and speak out plainly, I realized more keenly than ever how much more appropriately such an appeal to the common sense of the American people would be if made by my father or by some other of the little group of veteran packers who pioneered the industry. However, this being impossible, I put aside my own unwillingness, believing such a course to be dictated by my duty to the interests directly entrusted to my care, to the packing and private-car line industries in general, and finally to the American people.

The only color of propriety in my taking

up this unwilling work was to be had, it seemed to me, from the fact that the agitators in general, and the yellow magazines in particular, had selected the Armour house as their special point of attack as the *type* of all that they held to be objectionable. For this reason, and not because the heads of other packing and car line establishments could not better present the case of their industries, I have assumed the responsibility of attempting to present to the public of this country the falsity of the statements made by the sensation-mongers of the country who, for their own selfish purposes, have singled out the packing and the car line industries for a campaign of slander, abuse, and misrepresentation without precedent.

Let me disclaim at the start all superior motives, and admit that I am inspired by that kind and degree of selfishness which moves any man to protect his own, to be loyal to his responsibilities and to hit out against assailants who come at him with tactics that would not be tolerated in the prize-ring, on the foot-ball field, or in any form of contest where a fair fight is the accepted and enforced standard.

CONTENTS

ILLUSTRATIONS

THE PACKERS
THE PRIVATE-CAR LINES
AND THE PEOPLE

THE PACKERS
THE PRIVATE CAR LINES
AND THE PEOPLE

CHAPTER I

THE EVOLUTION OF THE PRIVATE FREIGHT CAR

A CLEAR and fair understanding of the private freight-car problem is wholly impossible without some knowledge of why and how the private car came into existence and how the system has been developed to its present proportions. Equally important to a right conclusion in this matter is a knowledge of the source and the animus of the present agitation, the avowed purpose of which is to put the private car out of business. Unless these things are clearly set forth the meat of the whole contention is missed.

But before I attempt to present these

essential facts let me say that I believe in the fairness of the American people, and that when once they get a clear understanding of a cause they will render a righteous judgment. President Roosevelt has put the disposition of this people most descriptively in declaring that they are "for a square deal." It is because I believe this so profoundly that I make the statements contained in this book. Any man who has a just cause need not, I believe, fear to appeal to the American people when he can be sure of reaching practically the whole people direct, and is also assured that he is not, in stating his case, liable to be misquoted or in any manner misrepresented.

The storm-centre of the present controversy is the private fruit and produce-car; consequently this must for the moment command the greater attention. The fact remains, however, that the meat-car was the pioneer in the private field, and that the fruit-car was a direct evolution from it.

CAR LINES AND THE PEOPLE

It has always seemed to me that the very manner in which the private car came into being is enough clearly to justify its existence in the mind of any fairly-disposed man of affairs.

In the old days cattle were moved from the Middle West to the East alive and on the hoof. This movement from the prairies, where the cattle could be most cheaply grown, to the centres of population in a part of the country ill-adapted to grazing, but where the demand for the meat was the strongest, was found to be very expensive, inland transportation then being both poor and costly. A steer weighing one thousand pounds would dress five hundred and fifty pounds. This is only another way of saying that shipping on the hoof meant paying high freight charges on four hundred and fifty pounds of what was *then* practically all waste material, and quite a large portion of the five hundred and fifty pounds of dressed meat also consisted of bone and other waste matter that could not be eaten. Of course, dressed meat could be shipped for a limited distance in freezing weather, but always with great risk, transportation then being slow, crude and uncertain. During the warmer months traffic in dressed meats was therefore confined to the local field.

Under these conditions it did not require any marvelous amount of acumen to see that the business could not be expanded beyond the local field, as far as the warmer months were concerned, unless some adequate means of protecting the meat against the effects of high temperature while in transit and until sold could be devised. Probably it is fair to say that the packers would have been decidedly obtuse if they had failed to see that the salvation and expansion of their business was locked up in the then undiscovered secret of how to ship dressed meats under some kind of cooling process that could be depended upon to keep them in good condition, and that should not be too expensive.

On the one hand, in the prairies of the West was the natural basis for almost unlimited expansion in the growing of cattle at the minimum of cost; on the other hand, the cities and the more thickly-settled sections of the East and South were calling for this meat from the prairies. The expense of shipping cattle from the West to the East was practically prohibitive so far as the natural growth of the cattle industry was concerned. The hour had struck for that great economic and industrial step, the appearance of the first crude meat refrigerator-car.

Properly speaking it was not a refrigerator-car at all; simply a big ice-box on wheels with the cargo of meat heaped upon the ice and coming directly in contact with it. Mr. Hammond, then of Detroit, is believed to have made the first fairly successful experiment in the building of a meat refrigerator-car, in 1871. Although stationary refrigerators were at that time in limited use, the facilities of the average eastern butcher or retailer for keeping meats were poor, and the western killers had, of course, no branch houses or distributing stations. The meat which came in contact with the ice became discolored and spoiled quickly when taken off and subjected to the inadequate refrigeration then almost universal. To remedy this difficulty cars were so equipped that the meat could be suspended from the rafters and ceiling, with the result that when the car was in motion and going around curves the halves of meat were set swinging like pendulums and finally communicated their motion to the car. Several wrecks occurred which were attributed to this cause, and the hostility of the railroads was awakened.

The next step in this evolutionary process was that of partitioning off one end of the car into an ice-bin, or bunker, the meat being suspended in the body of the car in a proper manner. This was a decided improvement, and the meat carried in these cars arrived in better condition. However, the true principle of car refrigeration eluded the experimenters for several years, but finally one of them hit it squarely and brought out the fact that a draft of air passing through a bunker, or ice-chest, in the upper corner of a car becomes chilled, so that it is heavier than the air which it meets, and consequently it drops down, circulating through the car, and finally, after it has lost its chill and becomes lighter than the incoming current, rises and passes out of the ventilator. Thus a current of fresh chilled air is constantly circulating about the meat, which is securely racked and does not touch the ice at all.

About the time this principle of car refrigeration was clearly established my father recognized its permanence and its relation to the future of the meat business. He saw that the meat refrigerator-car hitched the packing business to the growth of this country, that it annexed to the western ranges and prairie pastures the thickly-populated manufacturing and commercial centres of the East. Perhaps the other packers saw this as clearly as he did; certainly they were all interested in the development of the refrigerator-car. But he was tremendously in earnest in this matter, and went to the management of the railroads over which he would naturally ship the most meat and asked for the building of a small number of these cars. He was very promptly informed that this could not be done, that the demands for expenditures in the ordinary avenues were too great to justify going into a side issue of this kind. This brought him face to face with an emergency of almost critical importance. He saw that refrigerator-cars for the shipment of dressed meats were an absolute necessity, that their hour had arrived, and that the packer who did not recognize as inevitable this great change in the industry and make the most of it would drop behind in the irresistible movement of events.

On the other hand, his own business was, like that of the railroads, growing faster than his capital; he had hard work to keep up with it and needed more money in the routine expansion of the industry than he could command, without putting thousands of dollars into cars of the most expensive kind. All the vigor with which he could urge his case, his confidence in the great volume of business to be developed by the meat refrigerator-car, failed to move the railroad managers to whom he appealed, and he finally found himself absolutely forced as a matter of self-protection to the building of the first private car of the Armour system. To do this was then a decided hardship, an alternative which he adopted only because he saw it to be a matter of the sternest necessity. At that time the cars cost something more than twelve hundred dollars each, and because of the demand which he felt for money in the regular channels of his growing business he could build only a small portion of the cars which he really needed.

The meat refrigerator-car accomplished all that my father expected and more. Industrially it influenced the most important economic development of the packing business—the utilization of waste material in the manufacture of by-products. What this phase of the industry means to the people of the United States and to the packers I shall tell further on; but it is merely suggested in the statement that by its economies the packers are able to ship meat into thousands of localities remote from the great source of supply in the West and sell a superior "cut" at a lower price than the butcher can sell an inferior "cut" taken from a home-grown meat animal killed in his own slaughter-house.

But the refrigerator-car was destined to do more than revolutionize the meat industry. It was to place upon the tables of the people in cities and towns delicate and delicious fruits, berries, and vegetables grown in remote localities, as it had already brought to them the fresh meats from the prairies and ranges. Very early in the history of the meat refrigerator-car my father, who had been a farmer-boy in the fruit belt of New York state, became thoroughly convinced that refrigeration was the magic that would work as great a liberation and expansion for the fruit business as it had for the cattle industry.

After watching the results of experiments in shipping California fruits in ordinary meat refrigerator-cars he became so enthusiastic in his belief in what refrigeration would do in the handling of fruits that he called Mr. George B. Robbins (now president of the Armour Car Lines) into his office, discussed various points of difference in the requirements of a refrigerator-car adapted to carrying fruits and one for the shipping of meats, and finally said:

"Embody all these ideas in a plan and then place an order for the building of one thousand of the cars."

"But," cautiously suggested Mr. Robbins, "where are we going to get the business for so many cars?"

"Build them!" exclaimed my father in his hearty and decisive way. "I'll pay for 'em; then you do your best to run them right."

While the cars were being built men were sent out into the fruit-growing sections to hustle for business. When the growers of a district were first solicited to ship their fruits in a refrigerator-car they scouted the idea and declared that the fruit would be frosted and spoiled. In their minds refrigeration and freezing were synonymous, so far as the handling of fruits was concerned. Through sensible arguments and the testimony of growers who had once tried the experiment, sufficient business was secured to employ the new cars as they came out of the shops.

About this time my father took an extensive trip through the South and came back filled with the idea that it held wonderful possibilities for fruit-growers. At once he sent out a force of missionaries to see if the facts to be gained by a careful and scientific investigation would confirm this opinion. When the reports of these men came in he ordered another thousand fruit refrigerator-cars from the shops. These missionaries came into closest contact with the people and almost literally helped to plant and start one after another of the now famous fruit and berry districts of the South. These soon made themselves felt, and the third thousand of cars was soon under construction.

At length the pinch of hard times began to be felt throughout the country, and Mr. Mysenburg, of the Wells & French Car Works, confessed that if unexpected orders were not secured a shut-down seemed inevitable. My father immediately placed an order for two thousand more fruit-cars and advanced the money for their construction. From this time on he had a continued and increasing interest in the development of the fruit industry—a personal interest in the industry itself over and above that which he felt in the refrigeration business and its profits.

The growing of fruits and berries has been developed from the plane of comparatively an inconsequential avocation to the dignity of an immense industry. The number of private fruit refrigerator-cars in the Armour Lines has increased to twelve thousand. *The operation of the private fruit efrigerator-car has changed the growing f fruits and berries from a gamble to a usiness, from a local incident to a national ndustry,* bringing millions of dollars annu-lly to districts where land was worth only vo to ten dollars an acre before the general istribution of fruit was made possible by is agency.

It is quite natural, then, that the lay eader should ask: If the private car has one all this for the grower of fruit, why all is outcry from the fruit men against the rivate car? I am glad to have the ques-on raised, for the people, as a whole, do ot know the truth of the matter; it is time ey did, for they will be fair as soon as ey see the real situation fairly. Their resent views are based on misinformation nd malicious misrepresentations by those ho have an axe to grind, but who do not re to come into the open to do it.

The fact of the matter is that this whole itation started with the commission men of the country and not with the growers; these middlemen are the manipulators of the campaign that is being prosecuted for the express purpose of putting the private-car lines out of business. With Washington, D. C., as a centre, these commission men are pushing an extensive propaganda based on the cry that the grower is being oppressed by the private-car lines, and that this is the grower's fight for a chance to do business at a profit. All of the cunning at the command of these men is focused on the one purpose of spreading everywhere the impression that the private car is an "octopus" that is strangling the fruit-growing industry.

What are the facts in the case? The growers are satisfied with the private car, with its service, and with the system, which they recognize has been the biggest factor in building their business to its present proportion and stability and in opening to them the markets of the entire country.

This, I repeat, is not the fight of the growers, but of the commission men. The real attitude of the growers toward the private car was cleverly expressed in a speech by a representative of the Georgia Fruit-Growers' Association, who publicly declared:

"We have trained and chained the octopus so that it will feed out of our hands. The only thing we are afraid of is that this pounding of the refrigerator service by the commission merchant interests will cause the octopus to break its chains, jump the fence, and leave us, as in former times, *with no octopus* but *with all our peaches!*"

This is the sentiment of the growers everywhere. I do not believe there is one exclusive grower in this country who does not recognize that the private refrigerator-car is the salvation of and the mainstay of his business and absolutely vital to its prosperity and expansion. Also I as firmly believe that there is not one large grower in the country, having no interest in the fruit commission business, who does not clearly recognize that the private-car line, operating on the principle of the exclusive contract, is the only *practical plan* of handling fruit refrigeration, and that it gives the shipper a quality and reliability of service impossible under any other scheme of operation. Scores of fruit-growers' associations and hundreds of individual growers have given public testimony to that effect, and hundreds of pages of such testimony have been given in the form of legal evidence under oath.

Why, then, are the commission men so interested in putting the private refrigerator-car lines out of business? Because the private car has been steadily and irresistibly liberating the grower from the clutches of the commission man; because the private fruit refrigerator-car has compelled the commission man to quit doing business upon the capital of the growers and forced him to become an actual buyer and a merchant in fact. Under the old conditions of shipping fruit and berries the growers were absolutely at the mercy of the commission men. Fruit shipped without proper refrigeration stood a good chance of arriving in a more or less damaged condition, and this likelihood was the strategic stock in trade of the commission men, who were not slow to make the most of it.

Many reliable growers have testified that in the days before the private car invaded their territory they considered themselves lucky if the commission merchant did not demand a check from them to make up what the reported net results from the sale of their fruit fell short of the amount of the freight charges and of the middleman's commission. The consignments of fruit acknowledged by the consignees to have arrived in good condition were, in those days, about as rare as honest packers are now popularly thought to be. Or, if the grower's shipment was not reported to have arrived in damaged condition, he was likely to be told that the market was glutted, that the finest fruits were selling for what the poorest should bring, and that the shipper would do well if he did not have to send money to make up the freight.

Of course, it is true that without modern refrigeration great quantities of fruit did arrive at market in bad condition, and that a glutted market was altogether too common. But this situation was diligently and assiduously used by the commission merchants as a club over the growers' heads. A thousand pages could be filled with the evidence of growers who have suffered this sort of thing.

The coming of the private fruit refrigerator-car into a district put an end to this kind of tyranny. It carried the fruit of the growers into the usual market in precisely the same condition in which it left the orchards. There was no dodging this fact; it was so clear to the grower, the railroad and the commission man that it was beyond question or dispute. That did away with the time-worn excuse of the commission man that the shipment was received in badly damaged condition. The backbone of that stock claim was broken.

By the same token, the private refrigerator-car put the other stock excuse of a middleman out of service. The glutted market became a vanishing quantity under the ability of the private fruit refrigerator-car to take its cargo in prime condition to remote as well as near markets. In a word, not only did the private fruit refrigerator-car multiply the markets open to the grower and shipper, and insure the good condition of his fruit in transit, but the private-car system permitted him to divert his car at will and while in transit from its original destination to another less congested. In short, the private car enabled the grower to *know* that he could put his fruit into a *good market* in *good condition.*

Now these changes from the old line-up completely overturned the business of the commission merchant. The grower was no longer helpless in his hands, and the result was that the middleman had to get out and hustle for business; he had to go to the grower and treat with him on something like a fair basis. The history of every fruit district in which the private refrigerator-car has operated for a period of years shows that its advent was followed by the buyers, who went out to get their share of the crop and to bid against each other in order to do so.

It is clearly impossible to more than touch the high spots in a controversy of the magnitude of that which has thus been stirred up about the private-car lines—and to do so in merely a suggestive way at that. So far I have attempted little more than an introduction to the subject, but further on I shall deal in detail with facts and conditions.

Several of the most important phases of this fight must be indicated by at least a passing statement. One of these is the matter of rebates. The propaganda put out by the men who have set themselves the task of driving the private car off the tracks of the railroads of this country proclaims that the private refrigerator-car is simply a device contrived to evade the law forbidding the giving or receiving of rebates. This has been pounded into the ears of the people until, broadly speaking, they accept it without question.

So far as the Armour private cars are concerned, natural conditions, the force of circumstances, and the positive refusal of the railroad companies to put any money into "new-fangled refrigerator-cars" simply compelled my father to build them himself. He found that the refrigerator-car worked a wonderful expansion in the meat industry, and thought he saw that he could do as much for the fruit business. He backed his faith with his money, and pioneered the way in the development of the fruit, berry, and produce industry.

This development was not only natural but inevitable. Accident seemed to decree that the meat and fruit refrigerator-car should be builded by private enterprise, and certain industrial laws seemed to have further decreed that the private car should remain in private hands, and that its private operation and control give a far better service to the industries which employ it than could a refrigerator-car owned by a railroad. All this is very remote from a "device" artificially called into being to dodge the rebate law.

In fact, the Armour private car is not used as a device to secure, directly or indirectly, rebates, discriminations, or concessions to the car line owning it, to the shipper using it, to the individuals—or any one of them—owning the Armour Car Lines, or to any individual near or remotely connected with the industry. It was determined when the prohibitive law against rebates went into effect it was to be obeyed and not evaded; that a policy of indirection and evasion was a poor policy from any standpoint and would not be countenanced by the Armour interests.

That there is a considerable difference of opinion between the Interstate Commerce Commission and the railroads, relating to the transportation of property from interior cities of the United States upon a through tariff over railroads and steamships to foreign countries, is undeniable. The situation has provoked much comment. However, it is not clearly determined whether the matter is within the province of the Interstate Commerce Act and that of the Interstate Commerce Commission. The fact is that the recent indictment of railway and packing officials at Kansas City for alleged rebating relates exclusively to shipments from Kansas City to Europe. Speaking for my own company, the regular established public rates have been paid in full; and there has not been "any rebate, concession, or discrimination" of any character to shippers in this relation.

It seems to me that an "indictment" is neither the fairest nor the easiest method by which to arrive at an answer as to whether the railroads are required to file through tariff schedules relating to foreign shipments with the Interstate Commerce Commission or not. Shipments of other commodities, as, for instance, cotton, so I am informed, have always been made identically the same as the meat shipments involved in these indictments against the packers, but I do not recall any action being taken against the shippers of cotton. The manner of shipment in both cases has been fully understood by the Interstate Commerce Commission; there has been no concealment whatever in this respect.

This is purely a technical matter, a law

point which the Interstate Commerce Commission has long been urged to determine—one between the Interstate Commerce Commission and the railroads on jurisdictional grounds—whether the tariff sheet must be filed by the railroads with the Commission—a question that rightly involves neither the packers who are indicted with the railroads nor any other shippers to foreign markets. There is nothing in the Kansas City indictment that conflicts with my statement in the preceding paragraph.

Another element in the situation which has been subjected to bitter and sensational attack is that of the "exclusive contract." This also demands careful and detailed treatment, and will receive it later. Of course, this feature is inseparably associated with that of the reasonableness of the rate charged.

CHAPTER II

THE QUALITY OF THE SERVICE

IT should be explained, for the benefit of the general public, that the service for which the shipper pays the private-car line is that of refrigeration and not hauling. The private-car lines take care of the fruit from the moment it enters the car door until it arrives at its destination; it has nothing whatever to do with the freight charges.

Only the commonest kind of selfish common sense is required to arrive at the policy of keeping these refrigeration rates down to a point that will foster the fruit and berry industry and stimulate it to the broadest possible expansion. Any line of action less liberal than this would be short-sighted and suicidal. This is the policy that has steadily been pursued by the

Armour Car Lines, and no doubt by their competitors. The assertion that the rates charged for refrigeration of fruits by the Armour Car Lines have decreased in every district in which it has operated cannot be controverted. As the volume of business developed has increased, the rates charged have been voluntarily lowered. For example: When five different car lines were competing for the Georgia peach refrigeration business the rate was ninety dollars a car to New York. Afterward, under an exclusive contract with the Central of Georgia Railroad, it was reduced to eighty dollars a car. Again, in 1901, owing to an increase in the volume of business, and a slight decrease in the cost of ice, we voluntarily reduced the refrigeration charges to sixty-eight dollars and seventy-five cents.

The reductions made by the Armour Lines in the refrigeration fruit rates from California since 1895 vary from twenty-seven per cent. to fifty-five per cent., according to kind, loading-point, and destination. Everywhere throughout the whole field of operation the same consistent policy of reduction has been in force, and this has been brought out and substantiated in thousands of pages of evidence taken before the Interstate Commerce Commission, the United States Congressional Committees, and the other bodies which have had the private-car lines on the gridiron. The attitude of the grower is simply this:

"We want lower rates if we can get them—but, above all things, let nothing be done in an effort to get them that will interfere with or deprive us of the private-car-line service as it now stands, for we're prospering and growing under it and we can't afford to have it jeopardized."

When it comes to deciding whether refrigeration rates are reasonable or unreasonable we must consider the quality of the service, its cost to those furnishing it, and the hazards and liabilities which have to be shouldered by the car lines. That the service is indispensable to the grower, that he has paid far higher rates in the past than he now pays, and that his business has marvelously expanded under these rates, I have already indicated.

The quality of the service is so high that growers are of one voice in its praise. Broadly speaking, they have no complaints to make on this score. Of course there are individual complaints, but these are so insignificant as compared with the total number of shippers or the number of shippers who have put themselves on record as delighted with the service that my statement stands without qualification. Here is one incident, typical of scores of others demonstrating the growers' appreciation of the quality of the private-car service:

Benton Harbor, Michigan, is tapped by three railroads. Two of these, during the peach season just passed, furnished refrigerator-cars to shippers and did icing at the cost for ice of two dollars and fifty cents per ton. The private-car line operating over the other road charged its regular refrigeration rates. The apparent difference was great; but, in spite of this, sixty per cent. of the peach crop of that place was shipped by the private-car line, and eighty per cent. would have been carried if the road over which it operated had had engines enough to haul the trains.

A large proportion of the labor required in this service is above the ordinary in the matter of cost. Inspections, to be of any value, must be intelligent and dependable, and this kind of labor cannot be had at a cheap price. Private cars are inspected not only at the shipping-point, but at various points along the route.

Then let us look at the matter of hazards. Failure of the entire fruit crop in whole districts is one of the hazards. This happens altogether too frequently. It means, first, a total failure of revenue from that region, but that is only the beginning of the loss side of the account. Let me illustrate this by an actual occurrence representative of a routine feature of the business.

In 1898 the big ice-houses at Marshallville and Fort Valley were stored with ice to refrigerate the Georgia fruit crop; we had to ship that ice in vessels from Maine to Savannah and from there by rail to the inland peach country; this was expensive, but a big crop was expected the next season. A late frost wiped out the entire crop, and not one car of peaches was shipped out of the state. There was no way in which to use the ice and it melted. The ice hazard is one which catches the private-car lines both coming and going. A fruit crop that is abundant beyond all calculations often finds the local ice supply insufficient. There is only one thing to do, and that is always done. Ice enough is shipped from the nearest points (often quite remote), and this heavy addition of expense is on our shoulders and not on the shippers'; we get no more for the refrigeration because of this emergency outlay.

Under the exclusive contract the private-car line is obligated to do its part to have the cars on hand to handle the crop, and if it fails to do so it is responsible to the growers, and settles with them for the fruit lost because the refrigerator-cars were not there to take care of it. A very practical example of doing still more than this is had in an incident which cost our lines seventy-five thousand dollars.

Because of the extreme congestion of traffic on a certain railroad our refrigerator-cars were not at the shipping-point of the North Carolina strawberry districts at the critical moment. Therefore the berries could not be shipped before they became damaged. Did the growers suffer from this? Not at all. They received the market price for their crop, even a higher price

[illegible] Platform, Norfolk, Virginia

than if their berries had actually reached the market, for the market was short the number of car-loads for which we settled. We had contracted with the railroad tapping that territory to deliver so many cars to receive the ripened crop. Through no fault of our own the cars were not on hand, but we "made good."

There are many other reasons besides the one of liability pointed by this incident why the private car is the only logical agency by which the fruit business can be handled, and why the exclusive contract is the only logical basis upon which the private car can be operated. There is scarcely a railroad in this country operating in a fruit territory whose traffic officials have not testified under oath that it would be impractical, if not impossible, for their road to own fruit refrigerator-cars enough to take care of its own fruit business. Why? Because the peach season or the berry season, for example, lasts only three or four weeks; these cars cost over one thousand dollars each and are unsuitable for any other kind of traffic; the handling and care of them is a peculiar service which the railroads admit they are not equipped to perform; a railroad furnishing its own cars would not only have to furnish service along its own rails but beyond and wherever the cars might be.

Now, a private-car line doing business under an exclusive contract can accomplish practically all the essentials of good service which the railroads operating their own refrigerator-cars could not give short of a cost which would be absolutely prohibitive. It has a special and experienced service, its organization covers the entire United States, and wherever there is a natural highway for this kind of traffic there will be found its ice-houses or plants and its stations for re-icing and inspection. It commands the entrance to and the outlook over the markets of the country, and the grower has the advantage of this scope in every particular. He can send his fruit into any market and divert it *en route* if he desires.

As to the exclusive contract, it should be said that ice supplies, to be reliable, have to be stored up many months in advance of the fruit crop; cars have to be "parked" or concentrated long before they are used. Then an immense expenditure looking to the future has to be put out in ice-plants, other buildings, and for other equipment. Recently the Armour Car Lines put one hundred and twenty-five thousand dollars into an ice-plant at Las Vegas, Nevada, on the line of Senator Clark's new road that runs through the Mojave Desert and Death Valley, an arid and undeveloped region. Such an outlay would have been absolutely impossible and unwarranted if the car lines were debarred from making a long-time exclusive contract with the railroad looking toward and providing for the systematic development of a great fruit district and a great fruit traffic.

In almost every district where the exclusive contract prevails, at least so far as our lines are concerned, the growers themselves requested the railway to make an exclusive contract. They declared that under competition the supply of cars and the supply of ice was fluctuating and unreliable. No one knew just how many cars they would be called on for, and they had either too few or too many when the days of shipment came. Also they complained of the quality of the service under the competing system, declaring that it did not minister to the systematic development of the territory.

There are two or three features in the Armour system of distributing dressed meats which demand at least passing attention. I have shown, in detail, how the development of the modern refrigerator-car completely revolutionized the meat business of the world.

In accomplishing this wonderful transformation of food conditions, the refrigerator-car had a powerful and indispensable ally in the form of the packers' "branch houses," or distributing agencies, containing a reserve of fresh meats in the best of refrigeration, awaiting the call of the local retailers.

Not only does the branch house relieve the local butcher of the burden of providing extensive refrigeration facilities of his own, but it also allows him to carry a smaller stock than he would otherwise be able to carry without danger of disappointment to his customers.

Quite as important as either of these considerations is the fact that the local "branch" keeps at the demand of the retailers a supply of choice cuts, ripened to just the right point. At call, the retailer is able to go out and get for his most select and discriminating customers the best cuts in the best of condition.

No feature of the packing and dressed-meat business is more important than these branch houses—none more important to the public as well as to the packer. Neither expense nor attention to minutest detail is spared to make them models of what meat-houses should be, perfectly adapted to the purpose which they serve. They are built of the best materials that money can buy, and they are built on the best lines that ingenuity and experienced skill can contrive to secure perfect refrigeration and absolute cleanliness.

In the up-to-date branch house building-materials that are practically imperishable and impervious to outside influence are employed wherever possible. The floors are of cement. Storage and cooling rooms are lined with glazed tile, spotlessly white and smooth as glass. Not a cranny or crevice is left in which dust might gather or a germ hide from the frequent cleansings. All in all, they are as near dirt-proof, taint-proof and germ-proof as a building can be made.

These branch houses complete the packers' chain that takes the animal from farm or range, converts it into meat, and sets it down at the retail meat-merchant's door.

We would hear less criticism of the packers if consumers could follow a steer from pen to slaughter-house, see it converted into "quarters" and "cuts" and hung in the cooler, transferred thence to a clean, cold car, transported under ice to the farthest part of the country, and finally deposited in the branch house. That would bring home to the consumer, as nothing else can, the fact that no part of the people's food-supply receives more careful handling than does the meat that comes from the large packing-houses.

In the absence of such a comprehensive inspection it would be a distinct benefit to the packers if the general public would take pains to visit and scrutinize the branch houses. They are always open to visitors. If there is a branch house of Armour & Co. in your vicinity, you are cordially invited to see for yourself just how the hated packer takes care of your meat-supply.

The number of these branch houses maintained by the packers is very great. Armour & Co. have about three hundred of them in the United States alone. From the vast number of requests received from many sections of the country, asking for the installation of branch houses, it is very clear that these branches are looked upon by the people as being of great benefit to both the retailer and the consumer. Our aim is to protect the retailer and assist him in building up a secure and permanent trade.

In addition to the branch houses, we have, in many sections of the country, established smoke-houses. Pickled hams, etc., are sent green to these points and are there smoked under our own supervision, after which they are sent to the branch house. This enables the dealer to get freshly smoked meats.

One part of the system by which the packer distributes fresh meats to the people, through the local dealers, has been subjected to much criticism—and most unjustly, too. I refer to the "route car," by which meat is distributed to those towns not large enough to maintain a branch house or a distributing agency, or even to enable the retailer to order his meats in car-load lots.

The accusation is that the cars are used to "peddle" meats and thus hurt the business of the local butchers. It is not true that this is a peddling proposition. These cars are certainly of great advantage and benefit to the local butchers as well as the consumers, and were brought into existence to meet the present requirements of the community at large, and are not fairly to be considered as an advantage to the packers so much as an advantage to the people.

If the community were sufficiently large, car-load lots could be shipped to these various points to much greater advantage to the packer and at a considerable less expense. But the demand in the small place is as urgent as it is limited; the retailers and the consumers there must have fresh meats, but they cannot take them in large shipments; therefore the route car is indispensable to the people of the small towns.

It is far more expensive to ship in this way than to ship by the full car-load. Every time one of these cars is cut out of a train and put on a siding, the packer must pay from three to five dollars in addition to all the other transportation charges—and a car makes many such stops in the course of covering the route of small towns.

The question might be asked: Why not ship by local freight? Because meat would not arrive in good condition. Another reason for not shipping by local freight is that no dependence can be placed upon the arrival of meat thus shipped at a certain destination at any specific time. In short, this way is too slow and too unreliable for the transportation of fresh meats and meat-products.

The car-route salesman visits all the towns along his route and takes orders for shipments to be made on a specific day, stipulating that the car shall arrive at each place at a certain day and hour—to be met by the wagons of the retailers of that town. This method of delivery is carried out regularly once or twice a week, as occasion demands, insuring the consumer the delivery of his meats in the very best condition.

We do not sell to consumers, but reach them through the meat-dealers in the various towns, and our method of putting the meats in their hands enables them to get a fresh supply at very short notice, which could not be done without the route car.

Showing the practical working of this method of distribution, let me relate an instance. Mr. Boyd, formerly one of our branch-house managers at St. Louis, Missouri, is now extensively engaged in the retail meat business at Adrian, Michigan. From his wide experience in the branch-house meat business, he certainly knows whether it is now to his advantage, as a retail butcher, to secure his meats from these route cars.

Mr. Boyd is now getting the bulk of his beef-products from the route car running through his town, although he does, occasionally, go to the local butcher or slaughterer for some of his meats—but this only when he finds what he considers a "bargain."

Government inspection is another important feature of the packers' business.

To the general public, the meat-eating public, it ought to appeal as one of the *most important* features of any and all business in the whole country. It is the wall that stands between the meat-eating public and the sale of diseased meat.

This government inspection alone, if there were no other business or economic reasons, would be an all-sufficient reason for the existence of the packing and dressed-meat business on a mammoth scale. It should, if understood, make the general public a partisan supporter of the large packers.

Strangely enough, in view of its vital importance, this government inspection has been the subject of almost endless misrepresentation—of *ignorantly or maliciously false statements.*

The public has been told that meat animals and carcasses condemned as diseased are afterwards secretly made use of by the packers and sold to the public for food in the form of both dressed meats and canned meats.

Right here I desire to brand such statements as absolutely false as applied to the business of Armour & Co. I believe they are equally false as to all establishments in this country that are classed as packing-houses.

I repeat: In Armour & Co.'s business *not one atom of any condemned animal or carcass finds its way, directly or indirectly, from any source, into any food-product or food-ingredient.*

Every meat-animal and every carcass slaughtered in the Union Stock Yards, or stock-yards at any of the markets of the United States, is carefully inspected by the United States government. This inspection by the national government is supplemented, in practically all cases, by state or city inspection, or both. The live animals are inspected on the hoof and again when slaughtered.

The inspection by the United States government is not compulsory on the packers in the strict legal sense of the term; it is more binding than if it were compulsory. *It is business.* Attempt to evade it would be, from the purely commercial viewpoint, suicidal. *No packer can do an interstate or export business without government inspection.* Self-interest forces him to make use of it.

Self-interest likewise demands that he shall not receive meats or by-products from any small packer, either for export or other use, unless that small packer's plant is also "official," that is, under United States government inspection.

This inspection is carried on under the direction of the Bureau of Animal Industry of the Department of Agriculture. The packer has nothing to say about the employment of the inspectors. They are assigned by the United States government.

The government likewise is judge of their qualifications.

It requires of them, first, that they shall have taken a full three-year course in veterinary science—as long a course as most states require for the admission of physicians and surgeons to practice. Then these educated veterinarians are selected by rigid civil-service examination.

Every meat-animal that comes to the stock-yards is first inspected on the hoof, as stated, by representatives of the Bureau of Animal Industry. All that show signs of disease are segregated and tagged as rejected by the United States government inspectors. At regular intervals they are slaughtered (in Chicago under direction of the state officers) and consigned to other than food uses.

All carcasses—cattle, calves, sheep, hogs—are again rigidly inspected after slaughter. The internal organs affected by the various diseases to which meat-animals are subject are examined. On the slightest sign of disease the carcass is rejected, and so marked that it cannot escape observation.

From the moment it is rejected, that carcass is in the custody of the United States government agents and it is by them personally followed to the rendering-tank. It is hacked into small pieces, thrown into the tank, and emerges only as grease or fertilizing material. This tanking product is in such form that it could not by any possibility be renovated to become a food-product even if any packer were dishonest enough to attempt that.

And if it were possible to evade inspection and use condemned carcasses or product from an "unofficial" packing-house self-interest would again prevent it, because the packer would subject himself to speedy detection and exposure (if not endless blackmail) by the hundreds of employees who would be cognizant of his trickery.

This government inspection thus becomes an important adjunct of the packer's business from two viewpoints. It puts the stamp of legitimacy and honesty upon the packer's product, and so is to him a business necessity. To the public it is an *insurance* against the sale of diseased meats.

CAR LINES AND THE PEOPLE

CHAPTER III

MAGAZINE VS. ACTUAL PROFIT

IN actual business experience I have found nothing so immensely satisfactory as the profit percentages which certain writers are able to figure on the Armour Car Lines and packing business. The way these enterprising journalists can pile up paper profits for me reminds me of nothing so much as the net returns to be realized in the breeding of Belgian hares—on paper. Some of my readers may have had a little practical experience in breeding these creatures for market and trying to make the actual profits coincide with those so elaborately set forth in the prospectus sent out by the breeder of stock hares. In the prospectus the scale of profits is an ascending one in which nothing short of arithmetical progression is capable of computing the increase.

The net profits in the private car and the packing business are arrived at, by these entertaining writers, by the same process of figuring that the prospectus-maker used to determine the cumulative profits of raising Belgian hares. There is a world of difference between the actual earning of actual profits and the figuring of paper profits, where the total revenue is arrived at in a broad, generous and offhand way with a large ignorance and comprehensive disregard to costs, expenses, and all the practical elements and details really involved. If these writers could actually deliver the profit percentages which they are able to produce on paper they could command higher salaries in the packing or private-car business than any magazine in America can afford to pay them.

If any business man who reads, in these articles put out by the magazine writers, the statement of our profits is inclined to believe them, I have only this little act of justice to ask him:

Go to some neighbor who already has a profound conviction that you are making too much money, and let him figure, from hearsay information, the profits which you make. If you are then honestly content to be judged by a showing arrived at in such manner I will find no fault if you will accept as true the profits figured for the Armour Car Lines by these writers who start out with a necessity of making a case by showing exorbitant profits, and who are unhampered by information and have a splendid indifference to all matters of cost.

Just try this experiment; you will find it hugely entertaining. Perhaps also it will teach you how far you fall short, as a financial genius, of the expectations and ideas of those who feel that you are making more money than you have any business to make. This may be a little dampening to your pride right at the start, but it will suggest to you what you could accomplish if you could somehow contrive to eliminate the matter of expenses, and to use your best day's business as an "average" by which to multiply the business of the three hundred business days in the year.

The profits of the private-car business cannot, with any fairness, be judged on a harvest-time basis—which the critics of the enterprise seem to insist upon doing. It would be unfair to take the profit for a whole year and judge the business upon that basis. Why? Because this is a business of lean years as well as fat years. Then the period of profitable and established operation should be averaged with the years in which the business was in a struggling and pioneer stage. Absolute fairness would go further than this and take into account the exigencies of the future—such, for example, as the possibility that modern inventive genius may render practically useless and obsolete an equipment now representing an investment of millions of dollars, and this possibility is by no means so remote that good business prudence would not take it into account.

Again I urge each business man who attaches any weight to the profits of the Armour interests as they are figured in the magazines to act on this suggestion. The injustice of criticism and attack based on this kind of figuring will be so apparent to the man who tries this experiment that he will never again be tempted to place any confidence whatever in assertions arrived at by such a process. I deny that the profits of the private-car-line business and the packing business are extortionate. I have no desire to deny that both these interests do pay a profit. I should consider it a poor compliment to the manner in which I have discharged my responsibilities if this were not so. Let it be said, too, that the aggregate of these profits is respectable—

to many it probably looks immense; but it is not excessive or disproportionate when the immense volume of capital invested is considered.

Perhaps I am not called upon to say so, but I will make the statement that had I put my holdings, at the time I came into them, into railroads, national banks, and other enterprises, I should have made more money, made it with less trouble, and been subjected to less attack than I have been subjected to in the lines which I have followed.

More than that, I sincerely believe in the making of that money I should have been of far less service in the industrial development of this country than I have been in the private-car-line and the packing business.

Mind you, I am not posing as a philanthropist or asking for any credit on that or any kindred score. Just common-sense selfishness and the regard for the well-being of humanity that the ordinary decent citizen has are all the motives that I lay claim to in the conduct of my business; but I do confess to a sense of personal satisfaction in the fact that the prosperity of the property of the Armour Car Lines and the Armour Company has been inseparable from that of the fruit, produce, and cattle industries of the country; that thousands of men in these lines have been enabled to make independent fortunes by the activities of the private car; that, as an incident to a business success, the whole people of this country, and of other countries for that matter, enjoy comforts and luxuries otherwise impossible; that the standard of the world's living, in a sanitary and economic sense, has been immensely improved by reason of the operation of a business run for personal and selfish gain.

A reasonable amount of pride in these facts is, I believe, quite pardonable, and the satisfaction I get from this consideration is, I confess, quite as tangible and satisfactory a kind of dividend as I have been able to draw. I believe most profoundly that there are very few industries in existence that have contributed so much to the comfort and progress of this nation as the packing and private-car industries have contributed, and these benefits have been to all the people, for there are few who are not able to enjoy, to some extent, the fresh fruit, vegetables, and meat that the private refrigerator-car has guaranteed at all times to the public.

Too much emphasis cannot, in fairness, be placed upon the fact that the growers everywhere hail the private fruit-car as the direct agency of their prosperity and expansion, and that the cost of this service has certainly not been so great as to choke or stifle their development. Instead, they freely declare that the private car has liberated them from conditions which were choking and smothering their business.

Careless writers and persons with axes of their own to grind have sought to create the impression that "refrigeration service" as applied to fruit transportation is only a high-sounding synonym for "selling ice." Nothing could be further from the truth. Refrigeration service furnished by the private-car lines is exactly what the term implies. It is service—a *special* service—that insures to the grower or shipper of fruit a certain supply of the highest type of modern cars, careful loading, prompt moving, frequent inspection, and delivery of the fruit in good condition.

This service is comprehensive, highly organized, and expert. Transportation of fruit, care for a highly-perishable commodity, is its special and only work. Cost of ice is but one item in the expense of maintaining it. Its efficiency lies in the maintenance of a large and complex organization of trained men who are charged with the task not only of taking care of fruit in transit, but of seeing that the grower or shipper has cars, and the right kind of cars, in first-class condition, when he needs them, and not *about* when he needs them.

The charge for refrigeration service also covers maintenance, repair, and replacement of the tools employed in the business—the ice-making plants, ice-houses, icing-stations both in the fruit-growing sections and throughout the country along the routes from the growing locality to the market, repair-shops, and the cars themselves. These cars cost from one thousand dollars to twelve hundred dollars each—a third and sometimes half more than the ordinary box car. They are easily damaged. Almost every car needs more or less repairs every trip. They wear out more quickly than ordinary freight-cars, and are more easily put out of service from many causes. When returning empty, railroads frequently press them into use for other freight. If loaded with anything that leaves an odor—drugs, kerosene oil, etc., as happens often—the car is likely to be made useless for further service in the fruit-carrying trade.

To convey a concrete idea of what this fruit-refrigeration service means let us follow the shipping of a car-load of oranges from California to Boston.

California is a district in which a field organization must be maintained practically the year round. We have to maintain our own car-repair shops and icing-stations, and when fruit is moving a band of more than fifty men as inspectors, supervisors, etc., whose traveling expenses, as well as salaries, have to be paid, are up and down through the district superintending the loading and icing, enforcing prompt movement of cars, and pushing all details of the work. All told, we have a force of more than two hundred men in California during the season.

Before the fruit-shipping season opens, cars enough to handle the crop must be assembled at points convenient to the shipping-stations. More than half of them go West empty and are subject to many delays; some may be loaded with clean package freight. These will be from fifteen to thirty days *en route* from Chicago to Los Angeles, and unloading there may be delayed. During all this period, before a pound of fruit is loaded, the car line company must keep track of these cars, trace them from point to point, and know whether they are empty and available for immediate use or loaded and unavailable. Finally, the car we are following lands at our Los Angeles shop, where a large force of car repairers is employed at all times; they thoroughly overhaul the car, put new padding on doors and hatch-plugs, clean and repair tank-pans, drains and drain-traps, and attend to any other repairs needed, from supplying a new roof to a new set of trucks.

After the car is thoroughly overhauled and inspected to see that it is sweet and clean and in proper condition to carry a perishable load of fruit, its tanks are filled with about ten thousand pounds of ice. This initial icing alone in Los Angeles is a heavy expense. Wonderful things are grown in California, but no one, not even the wonderful Mr. Burbank, has been able to grow a natural crop of ice in southern California. For northern California we must haul the ice from the mountains at heavy expense for freight and shrinkage. We also buy enormous quantities of manufactured ice. In 1905 we bought more than one hundred and twenty thousand tons of ice in California.

When the car is iced it is sent out to the loading-point to receive its load. During this process there is a further heavy shrinkage of ice in the tanks. The car may have been standing for days in a railroad yard. The fruit loaded into it and the packages containing that fruit are also hot—soaked, as it were, in California sunshine. The car and its load must be brought down to a low temperature. Every board, every nail, every orange, every piece of wood, exudes heat. Much ice must be melted, obviously, to bring the hot car and its load down to a low temperature.

The car, when loaded, is sent back to Los Angeles (to Bakersfield if going East by the northern route), its ice tanks are refilled, it is thoroughly inspected again by car-line men, and is started on its journey East. If it takes the southern route it stops first at Tucson, Arizona, to be re-iced under the supervision of a car-line agent, who not only sees that the tanks are properly filled to capacity, but also makes a personal inspection of all drain-pipes, etc. The car then passes on to El Paso, where car-line agents are waiting for it. It is again thoroughly re-iced and inspected and sent on to Fort Worth, where another agent is in waiting to perform the same service. This process is repeated seven or eight times more before the car reaches Boston—Kansas City; Davenport, Iowa; Chicago; Galion, Ohio; Hornellsville, New York; and East Deerfield, Massachusetts.

The same careful attention is given it at each of these icing-stations as was given at the beginning of its journey. The same facilities are maintained on all the various routes. If slow time by the railroads or weather conditions necessitate a larger amount of ice than usual at any point, extra icing-stations are put in immediately. This is watched very carefully by an elaborate system in the Chicago office, and is also supervised by the traveling inspectors. They drop in on the various icing-stations unannounced. This work, however, is more of a precaution than a necessity. The agents who are strung across the country in nearly all cases have been with the car lines a long time, and are men who take a personal pride in the welfare of the business.

Let us not be understood that any of this frequent re-icing and re-inspection has been done at random. When the car leaves Los Angeles, the car number, its condition, and digest of the way-bill are all taken by a car-line inspector and filed with the district agent in charge of that territory. The car-line agent at the next station—Tucson in this case—is notified by mail, or by wire if necessary, that the car is on the way to him. This checking and notification of stations ahead continues without break until the car reaches destination. From the first icing until delivery at destination the car is under the eye of the car-line organization every hour and *is kept moving.*

When the car finally reaches Boston it is met again by a car-line inspector. He notes its condition, sees it opened, inspects condition of its load, and reports all details to the head office. This emphasizes the *insurance the service gives the shipper.* Nobody in Boston ventures to report to the shipper in California that a car handled and watched as described has "arrived in bad condition."

The history of this car as traced from California to Boston is typical. The same thing, generally speaking, happens to every private fruit refrigerator-car, no matter from what point it starts or to what point it is billed. Therefore, I repeat, the cost of this service embraces many items, to wit: Expensive ice in a hot, non-ice-bearing country, at the beginning of the journey, just where the most ice is needed; maintenance of repair-shops at various places (Chicago, Omaha, Kansas City, Fort Worth, Los Angeles, Sacramento, and many other points); salaries of inspectors wherever cars are iced; salaries of executive officers and clerks; cost of maintaining icing-stations; repairs; depreciation; interest on the investment, etc.

The item of repair is a heavy one. During a short, rush season, as in handling Michigan and Georgia peaches, time does not permit sending cars to the shops, but car-repairers have to be sent into the field. In Michigan last fall three car-repair superintendents were maintained in the field, and each had from five to ten men with him, at the expense of the car line. In view of these facts I submit that a tariff rate of seventy-seven dollars and fifty cents Los Angeles to Boston on the car outlined above is as low as good and proper service will permit. It was only a few years ago that the rate was ninety-five dollars, but we were able to reduce our operating-expenses and gave the shippers the benefit of it. Just as soon as conditions will warrant it the rate will again be reduced.

The private-car-line service also enables the shipper to control the destination of his product and to avoid glutted markets. It works this way: A shipper starts a car of peaches from Grand Rapids, Michigan, to Boston. After the car has left, he learns that the Boston market is full. At any place along the route of that car—Detroit, Buffalo, Albany—he can change its destination to New York, Providence, Philadelphia, Baltimore, or any other point that promises a better market. In 1904, during the one month of July, more than five hundred cars of Georgia peaches, an average of more than sixteen a day, were caught at Cincinnati alone, diverted from their original destination, and sent to other places that promised better results. This diversion of a car from one point to another is most *practicable with the kind of an organization maintained by the private-car lines.*

The point I desire to emphasize with these details is that the refrigerator-service tariff is not an arbitrary charge. It is based mathematically upon the service to be performed, and is *earned by the service rendered.* This view is held with practical unanimity by all growers and shippers who have had experience both with private-car-line service and with refrigeration service furnished by the railroads. One example of this will suffice.

In 1901 a certain railroad touching Benton Harbor, Michigan, was undertaking to furnish refrigeration service at cost of ice. Private cars were operating on another road reaching that same point. Mr. Roland Morrell, of Benton Harbor, one of the best fruit-growers in America, had twenty-five cars of choice peaches to ship. He was within three miles of a loading-station on the road which provided refrigeration at cost of ice. To reach a loading-station of the private-car lines his peaches had to be hauled five miles, part of the way uphill. Yet he turned his back upon the alleged low-price service, made the five-mile haul, shipped in private cars, and paid the tariff of fifty-five dollars a car to Boston rather than take a chance on railroad refrigeration service. Asked why he paid this "unnecessary" charge, he said: "I raise peaches to sell. I am not raising peaches to be spoiled in transit and paid for by the railroad."

Mr. John R. Wylie, of Shelby, Michigan, said to one of our representatives: "For long shipments we prefer the private-car-line service, with all its cost, to ice at actual cost with the imperfect service of the transportation companies."

At the close of the last Michigan peach season the *Fennville Herald*, which is the organ of one of Michigan's heaviest peach-shipping points, a paper that has never shown friendliness toward private-car lines, and is edited by the secretary of the Michigan State Horticultural Society, said that had it not been for the good work done by the Armour private-car lines in furnishing plenty of first-class cars and looking after the prompt icing of same in transit, that section would have been ruined the last season.

Almost every fruit and vegetable growing district in this country is a living witness to the pioneering work and the efficiency of the private-car line. These lines have served both to develop new fields and to widen the market of the fields already in existence when they entered business. This one fact alone should be sufficient to demonstrate the truth of that statement: there are practically no reliable statistics to be had anywhere in the country in relation to the fruit industry save those gathered by the private-car lines. A caller at the Agricultural Department in Washington, D. C., a few weeks ago asked for statistics on the fruit crop of Idaho. He was told that the department had none, and that aside from the apple crop the Agricultural Department had no fruit statistics whatever.

CAR LINES AND THE PEOPLE

Is it not remarkable that so many people should be eager to legislate definitely in relation to an industry that amounts to more than four hundred million dollars a year, yet on which there is so little information that the Agricultural Department of the government has no statistics whatever?

THE PACKERS, THE PRIVATE

CHAPTER IV

WHAT THE PRIVATE-CAR LINE HAS DONE

UP to ten years ago practically no fruit was shipped out of Michigan under refrigeration. Practically all of the crop was dumped into Chicago by boat and by ventilated cars. The Chicago market was uniformly low in consequence, and Chicago commission men made handsome profits by re-shipping Michigan peaches to other points, even back into Michigan.

The private-car line began to investigate the Michigan field some years ago. The car-line agent discovered that Michigan growers and shippers would not ship to eastern markets, such as Boston, New York, Pittsburg, etc., because they had no personal acquaintance with firms handling fruit at those points. The agent made it his business to get into communication with eastern fruit-dealers. Many of them were skeptical as to the statement that they could buy fancy peaches in Michigan. They were told to send their buyers into that district and, if they found that the results did not justify the effort, the car lines would pay the expense. Several of them took advantage of that offer; they came, were convinced, and bought.

This practice of sending buyers to the door of the grower—buyers who buy for cash and do not require the grower to ship on commission—has spread to all parts of the country. This is not the least of the advantages that the private-car lines have brought to the fruit-growers. *It gives the grower a market at his own door,* and his product is disposed of without risk to himself.

Since 1889 the fruit and vegetable industry in California has grown practically ten times in volume; and financially it is in better condition than at any previous time in its history. California shipped, in 1905, thirty thousand cars of lemons and oranges at an increase in profit over 1904 of more than one hundred dollars per car. At this writing two hundred cars a day are coming out of that state. The orange and lemon industry of California would not have been developed without the private car.

A few years ago head lettuce was a rarity in northern markets. The private-car line has developed this trade and has made many Florida farmers rich thereby. In the beginning, not more than half a dozen years ago, one car a day of head lettuce was sufficient to supply the New York market. New York alone now absorbs forty to fifty cars a day during the winter months.

A car-line agent interested strawberry-growers around Nashville, in 1903, to ship eight or ten cars as an experiment. Results were so good that the shipments rose to twenty cars in 1904 and to sixty cars in 1905. Humboldt, Tennessee, used to send out about fifty cars of tomatoes a season six or seven years ago. The tomatoes had to be shipped green and ripened in the commission man's storeroom, which, of course, impaired the quality. Humboldt now ships in a season five hundred cars of tomatoes that are allowed to ripen on the vines and therefore bring a much better price.

The new prune-plum district of Idaho has been developed entirely by the private-car-line missionary work and within a very few years. The far northwest now sends to market from two thousand to three thousand cars a year. Other important new districts are being similarly developed in northeast Texas, in Utah, Colorado, Arizona, Arkansas, and Missouri.

The principal fruit-growing districts of the country in 1899 shipped under refrigeration only nine thousand one hundred and sixty-four cars; the same districts in 1905 shipped forty-two thousand nine hundred and eighty-two cars. In particular districts during this period shipments have been multiplied to ten and even twenty times over so far as the Armour lines alone are concerned.

A phase of this development that is not to be minimized is the increased value given to the permanent investment in the fruit and vegetable growing lands. Michigan peach lands undeveloped are worth only from six to ten dollars and never more than twenty-five dollars an acre; with bearing peach orchards they command two hundred dollars to three hundred dollars an acre. Florida lands that were almost worthless are now yielding to growers of head lettuce and other early vegetables an annual return of five hundred dollars to one thousand dollars an acre. There are districts in Georgia where lands bought for one dollar an acre are now worth, with peach orchards in bearing, three hundred dollars an acre. California orange and lemon bearing lands are worth one thousand dollars an acre.

It is impossible to overemphasize the development work of the private-car lines in providing for the grower a *sure market* and a *profitable market.* The grower wants to know before he lays out money on his land that he will be able to deliver his products to markets in prime condition; the private-car-line service gives him that assurance. The shipper wants to know, when he loads a car of perishable fruit, that everything possible will be done to carry that fruit to any market or to the *best market* in good condition, so that it may command a fair price there; the private-car-line service gives him that assurance. The business of both the grower and the shipper is thus, as it were, *insured.*

That growers and shippers all over the country take this view is proclaimed in the reams of sworn testimony given in the past year or two before the Interstate Commerce Commission and in testimony before the Senate and House Committees of Congress at Washington. Mr. C. A. Sessions, of Shelby, Michigan, aptly summarized this phase of the question a few weeks ago in conversation with one of our agents. Mr. Sessions is essentially a grower—one of the most successful in the country—one who makes his peach orchard pay him all the way from fifty dollars to one hundred and twenty dollars an acre net profit.

"With private-car-line service," says Mr. Sessions, "buyers come to us—to our very doors—because they know this service will deliver in good condition what they buy. That makes a good market. Indifferent service keeps out the buyers. That makes a poor market and throws our fruit into Chicago, where the market is almost always glutted with fruit that goes across the lake by boat. Chicago commission men work to keep it glutted. We have had commission-house solicitors running up and down our streets here soliciting shipments, *when they knew Chicago was already over-*

stocked. That was to get our fancy peaches at a low, glutted market price, so they could be reshipped to other points at a profit.

"The effect of such a situation—growers deprived of efficient refrigerator-car service —is doubly bad. It depresses prices and it causes quality to deteriorate. In an overstocked market fancy peaches will not bring enough more than common ones to pay for the extra cost and labor put into growing high quality. One of the greatest benefits of the private-car service is that it has encouraged us to strive for quality. With that service, when we grow fancy peaches, we know we can get a good market for them."

Mr. J. R. Wylie, of Shelby, Michigan, supports Mr. Sessions with this testimony on Chicago as a market: "This last summer I shipped plums to two points, Chicago and Dayton, Ohio. The same kind of plums went to both places. My Chicago shipment averaged eighty cents a bushel net; my

Dayton shipment averaged one dollar and forty-five cents a bushel net."

A leaf from our own experience corroborates both Mr. Sessions and Mr. Wylie. Last fall representatives of two large eastern fruit-houses came West to buy Michigan peaches. They wanted to ship in private cars. At the Michigan points served by private cars they found competing buyers. They also discovered that quantities of Michigan peaches were being dumped into Chicago by boat and by rail from points not covered by the private-car service. Those agents forthwith came to Chicago, went on the open market in South Water Street, and there bought Michigan peaches for shipment East at lower prices than they would have had to pay over in Michigan at points served by private cars. More than four hundred cars were thus handled.

According to commission men of a certain kind—the kind who are the source of practically all agitation against the car lines—the private cars check rather than develop the fruit-growing industry. In view of what has been shown as to growers' and shippers' views it may be asked: "Have the commission men a motive?" Let us look at an example of what used to be a not infrequent experience of fruit-shippers before the guarantees of the private-car service were thrown around the business.

Not many seasons ago a grower in Georgia shipped two cars of peaches to an Indiana city. The consignee wired the grower that both cars arrived in "bad condition," intimating that the price would have to be cut. The grower asked the car-line agent's advice as to what he should do about it. He was advised: "Those cars left here in good condition and are, undoubtedly, in good condition now. Our reports will back up a lawsuit and probably enable you to collect for your peaches; but law is always expensive and you will probably save money by going yourself to Indiana now.''

The grower took that advice. His peaches had reached their destination on Friday; he did not reach there until Monday. Going as a buyer to the consignee firm (its members did not know him personally), he found some of his own peaches exposed for sale in fine condition.

''Got one hundred and fifty crates as good as those?'' he asked, indicating packages stenciled with his own name.

''Sure,'' was the prompt answer. ''We've got parts of two cars still on track—fine peaches all the way through. Come down and see 'em.''

The grower accompanied the commission man, saw his own peaches still in good condition after lying on a side track nearly three days, and then told his name. He was paid, without discount for bad condition.

A final word of testimony from the car lines' enemy—the *Fruit Trade Journal* is the organ of the commission men in their fight on the car lines. In the issue of September 23, 1905, forecasting the Michigan crop situation in a letter from St. Joseph, Michigan, that paper said:

"It is now evident that the Michigan peach crop will foot up six million bushels. . . . The marketing of fruit in southwestern Michigan has been completely revolutionized within the past few years. There was a time when the entire crop was thrust upon Chicago and Milwaukee, and the grower suffered materially from low prices resulting from a glutted market. Now the fruit belt, from the beginning of strawberries in June until the picking of the latest apples in October, is the *Mecca of fruit buyers from every part of the country, and the refrigerator-car* and steamboat do the rest."

In some quarters it has been made to

appear that the convenience of the public is deliberately defied and made to suffer through the operation of the private freight-cars. A cunning attempt to prejudice the public is made by the assertion that the railroads are so subservient to the Armour interests that they sidetrack passenger trains to let trains of these private cars pass; that the American citizen is held up on a switch so that the train of private freight-cars may have the right of way and not be interrupted in the work of earning mileage for a rapacious corporation.

Now it may be a fact that in some isolated instances local passenger trains have been sidetracked to let pass a through freight containing private cars. I do not know of such an instance, but it is possible that the exigencies of practical railroad operation might have brought this about in rare instances, but it is not a fact that this sort of occurrence is a part of the system of our operation or a logical result of it. Is there any reader of this book who frequently travels on local passenger trains who has not, at some time or other, been in a side-tracked coach that has been passed by a freight train made up of cars of miscellaneous kind and ownership? I think we have all had that experience. Certainly the experience is common enough to render absurd the insinuation that the railroads of this country are so dominated by the Armour or any other private-car lines or packing interests that they make a practice of giving trains of such cars precedence over their passenger service, and thereby subject the traveling public to delay, inconvenience, and indignity. Incidentally, it may be said that the slow passenger time from Chicago to New York is thirty hours and that the fastest "private-freight" time is sixty hours.

However, I make no denial of the fact that the private-car service, so far, at least, as the Armour lines are concerned—and I am willing to concede as much to competing lines—is so organized that its cars are not permitted to lag on the way, to loiter at division-points, or in any way to fail in delivering their cargoes at their destinations in the shortest possible time consistent with sound, safe, and reasonable railroad operation. In other words, energy, diligence, and perseverance are used in a systematic way to facilitate the transportation of fruits, produce, and meats as quickly and in as perfect condition as may be.

The perishable nature of the product demands "RUSH," and it is believed that this is distinctly a service to the grower, the shipper using the cars, and to the public buying the fresh fruits, vegetables, and meats carried in them—a service that needs no apology. If the "fast" fruit and meat-car service were allowed suddenly to lapse and fall back to the old-time running schedules, the result would be a public outcry and protest which would be shared in by the very people who are now sharpest in their criticism of the "fast" private-freight trains, and which would astonish the entire public.

Some critics of the private-car system are at great pains to create the impression that the mileage which the railroads pay the owners of the private cars as rental is so large that there should be no charge at all to the shipper for refrigeration. I will not ask the reader to accept my own statement that such a revenue would be a wholly inadequate compensation, but I will refer to the testimony of a practical railroad man not interested in the Armour car lines. Mr. J. S. Leeds, of the Santa Fé, made this statement under oath:

"The fact of the business is, the mileage that a car earns in the California fruit business will not maintain it. It will not pay the interest on its cost and pay for administration expenses of the organization and the replacements out of the mileage that it would earn. If this is true the refrigeration of these commodities should pay a profit and should also pay its share of the maintenance and the expense of the operation of the line."

It should be noted that the California business gives the longest haul, the most profitable mileage in the country.

Sensational periodicals have indulged in much and violent comment regarding the freight-rates given to the packers on dressed meats and other packing-house products as compared with the rates on live cattle, contending that when a rate on products as low or lower than that on live cattle is given a natural and fundamental law of rate-making is controverted.

More than this, it is contended that such a circumstance is *prima facie* evidence that the packers have a dominant and "monopolistic" power over the railroads and their freight-rates, and that those United States statutes known as the Interstate Commerce

Act and the Elkins Amendment have been violated. This matter was thoroughly tried out in the famous "Cattle Case" heard by Judge Bethea in the United States Circuit Court sitting at Chicago.

In his opinion, filed December 2, 1905, Judge Bethea, after a careful consideration of over one thousand pages of testimony taken before the Interstate Commerce Commission, and about three thousand pages of testimony taken in his own court, made the following findings of facts:

"That the live-stock rates are reasonable in themselves; these rates are equal to or less than the rates on dressed meats and packing-house products between the same points.

"That the cost of carrying live stock is greater than that of carrying dressed meats and packing-house products. In these cases, as to the particular commodities in question, the evidence shows that the defendant railroad companies pay out a much larger amount in damages for losses arising from the carriage of live stock than they do for losses arising from the carriage of dressed meats and packing-house products, in proportion to the value of the products carried, and more in damages per car regardless of the value. This makes the risk of carriage greater for live stock.

"The rates in question given to the packers at Missouri River and St. Paul were the result of competition.

"*That the competition in question did not result from agreement of the defendants, but was actual, genuine competition.*

"That the rates for carrying packers' products and dressed meats were remunerative.

"That the welfare of the public, including the shippers, consumers, and all localities and markets, does not seem to be materially affected by the present rates."

The essential finding of the court as to the

law was as follows: "The evidence above shows that Section 1 has not been violated —*the rates were not unreasonable.*"

Commenting upon this, the Chicago *Legal News* said: "Judge Bethea, in his opinion, declares that the *prima facie* case as made by the findings of the Interstate Commerce Commission has been overthrown by the evidence taken before him."

This whole contention could not have been more thoroughly thrashed out than in this trial, which lasted a month and involved examination of four thousand pages of testimony. The finding of the court should, it seems to me, most effectively settle this phase of the controversy, both as to the facts and the principles involved.

CHAPTER V

THE PACKERS AND THE CATTLEMEN

LET me make it clear that no one appreciates more than myself the fact that the cattle industry and the packing industry are inseparable in their fortunes. The one cannot prosper at the expense of the other; their interests are mutual, not to say identical. If the cattleman suffers the packer must suffer with him, and if the cattleman prospers the packer will naturally share in that prosperity.

Broadly speaking, the cattlemen understand this quite as well as do the packers. It is a truism to say that without cattle the packer could not do business—beef business. It is equally true that without the packing and dressed-beef establishments the cattle market would be small and unstable compared with what it is to-day. The progressive cattleman will not, I think, question this statement. And my purpose is not to tell the cattlemen that packers have done much for them—to boast of benefits conferred. Far from it. The packers have done only what progressive and enlightened self-interest has dictated, and, for one, I have no inclination to pose in the rôle of a benefactor. The cattlemen would be fully justified in sharply resenting any such attitude. This, I repeat, is entirely remote from my real feeling—quite as remote as any disposition on the part of the cattlemen to assume that they are the benefactors of the packers.

But the sensational magazines have persistently pounded it into the people at large that the cattlemen and the packers are in a perpetual state of warfare; that their interests are antagonistic; that loss or hardship to the cattleman must spell gain and prosperity to the packer; that the packer is a daylight robber, whose destiny is to prey upon the cattleman.

This malicious misrepresentation of facts and conditions is what demands a plain discussion of the relations between the cattlemen and the packers. The people should know whether the packers are robbing the cattlemen or whether they are pursuing an industry that is really the bottom and backbone of the cattle business. And the packers have a right that the public should understand the situation. Not even the possibility that some cattlemen may misconstrue the purpose of a plain statement of the advantages which the packing industry incidentally affords the cattle business should be longer permitted to stand as a bar against getting at the root of this matter.

Any fair-minded person who takes even a casual survey of the meat business will quickly recognize the fact that it was revolutionized and has been developed to its present immense proportion by the refrigerator-car, the modern system of canning meats, and by the scientific utilization of by-products—all of which are the fruits of the packers' ingenuity and enterprise.

One of the results of these elements has been the establishment of a cash market for every kind of cattle every business day of every year. There are some things which become so firmly established that familiarity with their routine operation has a tendency to cause their acceptance as a matter of course; they are so near and so commonplace to us, so unfailing in their operation, that we come to regard them as existing *by force of nature*—"by act of God" as the law puts it—that we lose sight of the fact that they were not always so from the beginning of things.

This, I think, is about the attitude of the average man towards a cash cattle market which is made possible by the operation of the great packing plants. He does not stop to think that there was a time when a steer might have been shipped to a market and without promptly finding a buyer—when, in fact, car-loads of cattle were shipped to market and could not be sold for cash; when there was no cash market ready to take the shippers' or the drovers' cattle at a going price and give him the money for them right on the call—and without regard to the kind or quality of his offerings.

Look at Chicago, the great central market: forty thousand cattle is not by any means a record day in the matter of receipts. Is it not a marvel that this enormous influx of cattle could be disposed of—and for cash? Is it not a wonder that they did not go begging and simply swamp the buyers? Could this task have been accomplished—or anything like it—had not the great packing-houses been here to utilize this monster herd and do it without delay? The official report of the Union

Stock Yards and Transit Company of Chicago states that this market handled, in 1905, cattle to the number of three million, four hundred and ten thousand, four hundred and sixty-nine, and to the value of one hundred and sixty-three million, nine hundred and forty-one thousand, six hundred and twelve dollars; also three hundred and eighty thousand, eight hundred and thirty-five calves, valued at three million, eight hundred and sixty-five thousand, four hundred and seventy-five dollars. It is true that forty per cent. of the cattle received were shipped out, but the remainder, going mainly to the packers, is so immense that the importance of the packing industry to this central cash market is too apparent to need argument. And remember that every market is a *spot cash* market. Hundreds of millions of dollars are invested by the packers in this industry, which is really a great manufacturing and distributing agent for the cattlemen.

Not only have the packers built up a great central cash market, but they have gone out to meet the cattlemen by establishing subsidiary markets in the heart of the cattle-country. These outposts of the packing industry have had their advantages to their owners, but I believe they have carried still greater advantages to the cattlemen.

One of the great advantages of the auxiliary market to the cattleman is the fact that it shortens his haul to market. This not only sometimes means a saving of freight, but the avoiding of shrinkage in weight and deterioration in quality. It also means that at the time of sharp demand he can get his cattle into the near market in time to realize the high price, while he would not, perhaps, be able to rush them into the distant central market before the extraordinary demand would be satisfied and prices drop back again.

In the big central market there is naturally a larger call for cattle for export

A Bunch of Steers for Armour and Company's Packing House, Chicago, Illinois

and for the fancy trade demanding choice cuts, but it is a fact that all grades bring, in the auxiliary markets, prices as high as they do in the central market.

Another consideration not to be overlooked in this connection is the fact that the subsidiary market has immensely influenced the general production of a better quality of beef by facilitating the feeding or "finishing" of cattle brought in from the ranges. Named in about the order of their establishing, the principal subsidiary markets developed by the packers are: Kansas City, South Omaha, East St. Louis, St. Joseph, Sioux City, South St. Paul, and Fort Worth. The extension of the packing industry to these points has changed the agricultural map of the states tributary to these auxiliary markets, making them the richest feeding-grounds in the country.

The farmers of these regions go into the near-by markets and pick up herds of cattle brought in from the range, take them to their barns and pastures, and subject them to a finishing process which puts them into a much higher class, as beef animals, than that in which they belonged when they left the range.

Also the outpost plants of the packing industry and the markets created and maintained by them have immensely stimulated the breeding of fine high-grade cattle. There is scarcely a locality in the Middle West where the leading farmer has not his herd of fine Herefords, Shorthorns, Aberdeen-Angus, or Galloways, and who is not striving to increase his income by *improving the quality of his cattle.* This means the placing of more and more good beef on the market year after year.

Right here is the place in which to note the effect which the auxiliary markets—and the central market, too, for that matter—have upon the growers of corn. The feeding of beef-cattle in states tributary to these markets has greatly influenced the corn market.

B. W. Snow, the well-known crop expert, is authority for the statement that more than eighty per cent of the corn crop of this country is consumed practically upon the farms where it is grown. Or, to put it more exactly, less than twenty per cent. of America's corn crop is moved out of the county in which it is produced. When it is remembered that the average corn crop of the United States is two and one-half billion bushels annually, the extent of our feeding operations may be dimly realized—but only dimly, for the figures go beyond the scope of the imagination of most of us.

What it means to feed more than two billion bushels of corn right on the farm producing the crop is not to be grasped without some analysis and consideration. For one thing, it means that the farmers who grow and feed this vast volume of corn are manufacturers who turn out a finished product, and by so doing, get the advantages and benefits of by-product utilization. This is just what the packer does; if he did not do it, his annual balance sheet would make a sorry showing under present conditions.

By feeding his corn right on the ground where it is grown, the farmer retains for the enrichment of his land the benefit resulting from the animal digestion of his crop—a value amounting in the aggregate to millions of dollars a year. The vast extent of the feeding business is a substantial testimony to its profitableness—a testimony so convincing that additional evidence is scarcely required. Only those unfamiliar with the subject or who come to it with a prejudice will dispute the flat statement that, generally speaking, the growing and feeding of corn is profitable and that the main element in the prosperity of this branch of agriculture is the packing industry. If the business of killing cattle, hogs and sheep was in the undeveloped stage in which it was when the evolution of the packing industry began, would our western farmers be growing two-and-a-half billion bushels of corn a year and feeding to their own stock more than eighty per cent. of that crop? I do not think any reasonable and intelligent person will say that such would be the case.

The fact is, that nothing short of the vast and highly organized packing industry as it stands to-day, with its immense capital and its immense and constant demand for raw material, could sustain so great an agricultural business as that of the combined corn-growing and corn-feeding industry. In other words, if the farmer could not sell his stock any day in the year and for cash, the present development of corn growing and corn feeding would be impossible.

When it comes to the matter of our foreign trade, the dependence of the corn-grower upon the packing industry is most clearly apparent. It is not too much to say that our large export trade in dressed meats and meat products is *due* wholly to the packers; but of this I shall speak in greater detail elsewhere.

What kind of farm lands has shown the greatest advance in the last fifteen years—the period in which the packing industry has been mainly developed?

Corn lands! In that time the average value of corn lands west of the Allegheny Mountains and north of the Ohio River has doubled, broadly speaking. In Illinois, fifteen years ago, sixty dollars an acre was a good price for corn land; to-day the ruling price is from one hundred and twenty-five to one hundred and seventy-five dollars. In Iowa, Nebraska and eastern Kansas the proportionate advance has averaged still greater.

Every year this nation devotes eighty million acres to corn. Our national cornfield is larger than many a European kingdom which looks big in the eyes of the world. This fact, it seems to me, is a clear indication that the business of manufacturing meat on the farm is a profitable one and that the packing industry has fulfilled a useful service in furnishing the distributing organization for putting this product into the markets of the world.

If the prop of the feeding business, which rests upon the packing industry, were withdrawn, the corn-producing states would be dealt a blow which would send prices down to the low figures which prevailed before the feeding business was developed by the packing industry—prices so low that they now sound strange and almost unbelievable. The foundation upon which the great cash corn market rests is the feeding of cattle and hogs, and this great branch of modern agriculture is almost wholly dependent upon the packing industry, as that term is understood in its broadest sense. Strike out the business of feeding, and the farmer who raises corn would stagger under the weight of an almost hopeless calamity.

The cry of "monopoly" and "combine" is not new—not even in the meat and cattle industry. Not a man would to-day deny that the establishment of the Chicago Union Stock Yards, taking the place of a half-dozen yards scattered all over Chicago, was one of the best things that ever happened to the live-stock industry. The Union Stock Yards were opened Christmas Day, 1865. Less than a year later—in November, 1866—the *Prairie Farmer* of Chicago had scented a "combine," and proceeded to expose it in these words:

"A mischievous combination of buyers and sellers to prevent the producer from learning correct values has destroyed public confidence in great measure.

"There is no disguising of the fact that through the manipulation of a few unprincipled buyers and sellers styling themselves 'the board,' the prestige that Chicago has heretofore enjoyed of being the great live-stock centre of the great northwest is fast slipping from her grasp.

"A majority of dealers in the country look upon the great Union Stock Yards as a market where swindlers and live-stock shysters 'most do congregate.' A reform is needed. The public demands it, and unless this demand is heeded the combination on the Union Stock Yards will find their occupation gone."

Yet the live-stock industry and the Union Stock Yards waxed greater in spite of the "combine," handling three million, four hundred and ten thousand, four hundred and sixty-nine head of cattle, worth one hundred and sixty-three million, nine hundred and forty-one thousand, six hundred and twelve dollars in 1905, as against four hundred and three thousand, one hundred and two head in 1869.

CHAPTER VI

A CAMPAIGN OF SLANDER

THE sensational periodicals have the advantage of the packers in their campaign of slander. Human nature is such that an attack on an individual or an institution always helps to sell the periodical containing it, but the packers cannot sell their meats by abusing these publications which assail them. These magazines have the further advantage of us in their peculiar character. They are constructed not for the careful, but for the careless reader. They are written and edited with one thought in mind—to produce an effect, to make a point quickly, to leave with the hasty reader an impression that will forward the purpose of the publisher, whether that purpose be political, sociological, or commercial. Hence it comes that the writers and editors of these magazines have borrowed from the stage a form of dramatic license, and dress up their literary merchandise to *produce an effect upon the reader,* just as the playwright exaggerates his situations and the actor resorts to the exaggerations of "make-up" to produce an effect. These assailants of the packing industry have not hesitated to deal in half-truths and "cooked-up" evidence, and to distort and juggle plain facts into absolute falsehoods.

One magazine writer employed a series of cartoons—a diminishing series of pictures of a steer—to show how the price of the cattleman's stock has been crowded down by the "trust" since 1900. He carefully omitted a representation of prices in 1902, when cattle-prices were at the highest point ever reached in twenty years.

The same writer, with much use of black-face type and capital letters, set forth that "forty Iowa banks were forced to close their doors'' in 1903-04. The statement was so framed as to appear, on hasty reading, as if the Iowa State Auditor's office was authority for the whole statement instead of only that part of it which gave the names of the Iowa banks that had failed within a certain period. When this statement was called to the attention of the Iowa State Auditor's office, Chief Clerk Cox, of the Banking Department, denounced it as utterly untrue, tabulated the list of banks and gave the reasons for each failure, which reasons all came under the head of unwise speculations and reckless banking methods. Another magazine writer stated—to bolster up the allegation that the ''Beef Trust'' forces the railroads to discriminate against others and in favor of the ''trust''—that beef, the high-priced product, is shipped at eighteen and a half cents, and cattle, a low-priced product, shipped at twenty-three and a half cents. Movement of live stock and its product, or packing-house products, from the Missouri River points to Chicago is referred to.

The above statement, though technically true, is entirely misleading. While some of the railroads publish a local rate of twenty-three and a half cents on live stock from Missouri River to Chicago, less than one per cent. or practically *none* of the business is moved under this rate. Live stock from Missouri River points come in there from the West, and all of such business, when reshipped to Chicago, takes the proportional rate, varying from thirteen to seventeen cents per hundredweight from Missouri River to Chicago, so that, in fact, the live stock is carried at a less rate per hundredweight than the product.

The local rate on the "product" from Missouri River to Chicago is also twenty cents per hundredweight instead of eighteen and a half cents, although the proportional rate on the product from the Missouri River to Chicago on shipments consigned through to eastern points is eighteen and a half cents.

One more incident: When known corn conditions foreshadowed very high prices for cattle, a Chicago paper printed an analysis of conditions and a forecast of consequences — higher-priced cattle and higher-priced beef—as a part of its regular market report away in the back pages of the paper. That article was written by a *market expert* as a reflection of *market facts,* to be read and judged by *men in the business,* most of whom *knew the facts.* A little later, when the conditions outlined in the market report were commencing to work and prices began to go up, the same paper printed a sensational article *on its first page* relating how the "Beef Trust" was putting up the prices of the poor man's food.

The dressed-beef and packing industry is not in the hands of a "monopoly combine," never has been in the hands of such a "combine," and never will be. *It cannot become a "monopoly."*

The business of the packers is not sheltered by tariff nor builded upon patents of secret processes. Their raw material is not to be gathered from the bosom of the earth at no cost beyond the mere expense of extracting it. Armour & Co. do not own or control the sources of their raw material, and are not even interested a dollar's worth in the production of raw material, and I do not think any of the packers are so interested to any extent. They do not own or control the transportation avenues over which the raw material comes to market. They do not, or could not if they would, control the means of distribution to the consumer, because this product goes, not to comparatively a few large users, but to each family individually that helps to make up the total of millions on millions of eaters of American meat on this continent and throughout the world.

Without control of some, or at least of one, of the commercial instrumentalities mentioned no industry can be monopolized. Consider a further bar against monopolizing it: the kind of food it deals in can be produced, made merchantable, and distributed to the consumer in each state, each county, and each township of the entire United States—and many other countries as well. Every farmer cannot find an oil-well or a deposit of iron-ore in his back pasture-lot, because Nature has planted them only at rare intervals, and he cannot grow sugar-beets in every field, because sugar-beets require a peculiar character of soil; but every person who owns a little land that is not absolutely barren can engage, to some degree, in meat production, and almost every man who has a knife and a saw can engage in the slaughtering business.

In the language of the United States Census report of 1900: "The process of converting live stock into food for human consumption is an industry which, directly and indirectly, furnishes employment to a considerable portion of the United States, and sustenance to all.''

The industry of the packers is dependent upon supply and demand to a degree that prevails in no other large industry. If it were of less magnitude it might be fittingly called a hand-to-mouth business. It is, broadly speaking, a manufacturing business, and one that in some respects is more hazardous than any other.

The general run of manufacturers of any considerable size handle raw materials which are not perishable in the usual sense of the term. Even when their raw materials are perishable they become less so, if not practically imperishable, as soon as passed through the manufacturing process.

The packer uses a material that is quite perishable, and a large part of his finished product—fresh meats—is highly perishable; so he is *taking risk at both ends of his business.* He cannot "scalp the market" or "discount the future," take advantage of a temporary market condition and load up with raw material to be manufactured at his leisure and at a great profit; nor can he, when the demand slumps, continue to run his plant and store up finished product against the day of higher prices.

The "frozen cuts" of beef furnish the only exception to this statement, and they amount to only about two per cent. of the finished product. And this frozen-cut business is of distinct advantage to the cattleman because it enables the packers to take care of the immense floods of cattle which come in from the ranges in the fall of the year. From day to day, almost literally, his purchases of raw material and his sale of finished product must balance. His profit, if he make any, must come from stopping every leak, saving waste, and turning his money over rapidly at a small margin of profit on a large volume of business.

The industry exemplifies the ideal business theory of "quick returns and small profit." It gathers the product of the millions of small producers throughout the Western Empire stretching from the Alleghenies to the Rocky Mountains, converts their product into merchantable commodities, and distributes them to the consumers of the whole round world. For the service it performs it is none too well paid in the profit it makes—an average of less than two per cent. on the volume of business handled.

To all practical purposes, the packer is the agent of the cattleman, handling his cattle, as the official report of Commissioner Garfield says, for an average fee of ninety-nine cents a head. This vast distributing agency is as readily at the command of the man who has a herd of ten or a hundred cattle as the man whose herd numbers a hundred thousand—and on precisely the same terms, too. But in one respect this way of looking at the packer as agent or commission man acting for the cattleman is inadequate—for the packer pays cash every day in the year. He does business strictly upon his own capital and pays "at the drop of the whip."

The character of the packing business, I repeat, and the wide distribution of it preclude making it the property of a monopoly. The business methods, practices, and necessities pertaining to it are a further bar against monopoly.

On one side stands the cattle-grower. He has absolute control of his product. He can ship to market to-day if he will, or he can wait a day, a week, a month, in ordinary cases, without much risk.

The packer looks to him for raw material —live stock—and cannot get it until it comes to market.

On the other side stands the retailer of

meats. His is a from-day-to-day business. He buys only as the demand tells him to buy. The packer must market his finished product through the retailer. He cannot force or induce him to buy one pound more than he wants to buy. Between these two commercial factors stands the packer. *He must do all the guessing at both ends of the line.* If he does not buy cattle fast enough, the demand from the retailer outruns his supply and he loses the business. If he buys too many cattle he must hold them at heavy expense (for he is without facilities for storing cattle) or convert them into meat for which there is no demand, and run the risk of having it spoil on his hands.

It is up to the packer correctly to judge the balance between the supply and the demand. Thus each day's business becomes a separate business. He cannot know, his buyers cannot know, when the cattle market opens on any given morning, what the market for that day will be or ought to be. He and his buyers may know, in a general way, what the market and prospects for meats are, based on the day before, but each day's cattle market is a new market with a new lot of cattle, resembling in no way, perhaps, the cattle of the day before.

No combination that could be formed would serve to keep a "Beef Trust" advised of the character or number of cattle coming into market on any one day from all points of the compass. Yesterday may have brought a heavy run of choice beeves. To-day's receipts may be common. To-morrow cows may predominate. The packers have no advance information, and their cattle-buyers have only general instructions as to the needs of the house for that day. They exercise their wits and their judgment, buy as closely and as carefully as they can, drawing the line between buying too much and too little. They draw that line very fine. Their jobs depend upon it, and that is the whole mystery of cattle-buying.

The cattle market at Chicago, Kansas City, South Omaha, and everywhere else *is an open market.* Every buyer of cattle in a market, whether buying for a packer, for reshipment East or across the water, or for his individual account as a feeder or speculator, stands on absolutely even terms with every other buyer. The competition is such that almost every good bunch of cattle will draw out several bids, instead of the *one* bid the "trust" exploiters allege.

Large packers will not deny that they can slaughter, pack, work up by-products, and sell to better advantage than some of their smaller competitors. Each packer thinks he can do all that better than any other packer. If he did not think so there would be little excuse for his remaining in the business. Every merchant thinks he is, in some particular, a better merchant than his competitor. That is why he is a merchant.

If a packer ever buys cattle to better advantage than his competitor, it is because his cattle-buyer is a better judge of cattle than his competitor's buyer, and that brings in the personal equation—one man's brains and judgment against another man's—and that cannot be governed by a monopoly.

The cattle market, I repeat, is an open market, and no person who has even a chance acquaintance with a cattle market in action will venture to suggest that any packer or group of packers can "take his pick" or buy any particular bunch of cattle cheaper than the smaller buyer in the market. If that were to happen you would hear a roar of protests from buyers and from cattlemen alongside which the present cry against the "Beef Trust" would be tame and flat.

There is no cleaner competition anywhere than among the cattle-buyers on the live-stock market. *This competition,* together with the marketing methods followed, *automatically regulates the market,* keeps it an open market, and prevents control of it by any "combine." To make this clear, let us look into the ways this selling and buying of cattle is carried on.

The buyers in the market at the Union Stock Yards of Chicago—leaving out buyers of feeder cattle—may be classed as buyers for the large packing-houses, buyers for the smaller packers and slaughterers, buyers for shipment to the seaboard and to the Old World, buyers for speculators. These different classes of buyers have nothing in common. They are always at war, commercially speaking.

CHAPTER VII

CATTLE AND CATTLE MARKETS

THE packers are always in the market for cattle. They have large plants, which if allowed to lie idle do so at great loss. As long as they pay the top price—always with an eye to the selling market for the product on the other side of the slaughter-house—they get their choice of the cattle. But if buyers for the large packers should combine to depress prices, what would happen? The moment prices went to a point that promised a little extra profit on the slaughtered product, the buyer for small packers, for shipment, and for speculation, would sweep the market, and back prices would go over the heads of a badly rattled "combine."

There may be an impression that the buying capacity of all except representatives of the large packers is too limited to have much effect on the market. The best answer to that lies in the official figures. In 1904 the receipts of cattle at the Chicago Union Stock Yards were three million, two hundred and fifty-nine thousand, one hundred and eighty-five head; of these one million, three hundred and twenty-six thousand, three hundred and thirty-two head were reshipped—to feeders, the seaboard packers, to Buffalo, Cleveland, Cincinnati, Pittsburg, Indianapolis, other cities, and to Europe. During 1905 receipts were three million, one hundred and eleven thousand and twenty-nine, and the shipments were one million, four hundred and five thousand, seven hundred and eight. Thus more than forty per cent. of the cattle received were bought for shipment.

It is important to remember in this connection the fact that there is not a slaughtering establishment in this country that is regularly run at its full capacity. See what this means so far as the control of the cattle market by any "combine" is concerned. The instant cattle prices became depressed so that there was an attractive margin, the smaller packers and killers would jump in, get the cattle, and kill extensively. There are hundreds of these smaller houses which make a business of waiting upon the turns of the market for the hour of opportunity, killing at certain times two, three, and four times as many cattle as they do when prices rule above a certain low point.

If the packers could regulate the cattle market they would prefer to have a steady market with an even inflow of cattle—about the same number each day. Their profits depend upon the rapid turning of the money invested, upon shortening as much as possible the period between the moment when steers arrive and the time when their beef is sold. When their plants lie idle they lose. Every manufacturer's aim is to keep his plant in even and continuous operation.

Right here it should be remembered that the large packer must have, every day, a certain amount of high-grade cattle. The only way he can get this is to go into the market when it opens. If he were to hang back he would get left on this imperative material; he would get only the "tail ends." His only protection in this particular is to buy early. The sort of buying methods attributed to "trust buyers" would leave him in the lurch on this score.

An important chapter in any comprehensive history of the development of the cattle business would be the chapter on cattle loan companies. These companies assisted to develop and stimulate and make a business of cattle-raising in a measure difficult to overestimate. Among the first one of these was the Omaha Cattle Loan Company, organized nine years ago by Thomas B. McPherson. The packers backed it with their money and credit as they backed others at Omaha, Kansas City,

St. Joseph, and Chicago—for selfish reasons, of course.

These loan companies, managed by men who knew cattle, made a specialty of loaning money on cattle, and thus put life into the industry. Before their time, cattle-raising on a large scale was practically closed to the man of small capital. The local banks, where there were banks, were too weak to take the risk; but with the cattle loan companies in business the capable cattlemen were able to go into the business on a large scale. These companies helped develop the growing of higher-grade stock. An idea of the importance of this feature may be gathered from the fact that the Omaha concern loaned ten million dollars annually almost entirely on ranch and feeder cattle.

True, these companies are all out of business now—put out by two causes. Enforcement of the no-fence law by the United States government has made loaning on large herds extra hazardous. Then the prosperity of the West, to which the cattle business, stimulated by the cattle loan companies, has contributed no small share, has filled the local banks with money and has enabled the local banker to loan the money on cattle that used to come from these cattle loan companies.

Now consider recent low prices. Natural causes have continued to hold cattle prices down with the break of 1903. Chief among these causes has been the breaking up of the big western and northwestern range herds consequent upon enforcement of the no-fence law by the United States government.

Since the buffalo gave way to the steer on the western plains in the late seventies, a considerable proportion of the beef-cattle supply has come from the western ranges. As the packing industry developed and, by making an assured market, converted range-cattle-raising from an adventure into a settled business, the cattlemen learned that it was better to fence the ranges than to herd the cattle on the open plains.

All around him in Montana, western Nebraska, Wyoming, Colorado, Idaho, and the Dakotas were millions of acres of grass-land fit for nothing but cattle-grazing, and fit for that only where water could be had. By acquiring ownership of a small tract surrounding or adjacent to a water-supply, he could practically control thousands or millions of acres surrounding him. If it was public land he used it for nothing. If it was Indian Reservation land he rented it for a cent or a fraction of a cent an acre a year. With a barbed-wire fence he could inclose what he needed and go into the cattle business on a large scale.

A few years ago the United States government began to enforce the law against fencing these lands. The range cattle-man has not found and cannot find a substitute for the fenced range of which he has been dispossessed. Without fences his herds cannot be kept together. This increase in the hazard of the range-cattle business has made this class of cattle a poorer loan risk. Loans have been withdrawn, thus imposing an additional burden upon the cattleman. Therefore he has bowed to the inevitable and has broken up his herds.

Some of the cattle from range herds so dispersed have found a market among western feeders, but the bulk of them have come to market and have been sent to the slaughter-houses. Then, too, the wonderful abundance of grass helped to make 1905 almost a record year in the number of cattle sent to market. The West, from Texas to Canada, has been literally a garden as to pasturage.

These cattle have been a weight on the market for the past two years. During 1905 three hundred and eighty thousand head of range cattle were marketed in Chicago alone—a larger number than was ever received in this market during a similar period except in 1894. At all the market points for western cattle the receipts will total about three-quarters of a million head.

Very many of the big range herds have already been broken up. Nineteen hundred and six will, I think, see fewer range cattle in the market than in 1905, and the year after still fewer. If one were in a prophetic mood he might say that these conditions will produce, in a few years, much higher-priced cattle and consequently higher-priced beef. We shall continue to raise beef-cattle in this country, but at greater expense.

It has taken, on the range, ten to twenty acres to produce a steer, but these acres cost practically nothing, and the steers made good beef. Sometimes they competed with choice fed steers for the export trade. Fewer acres will produce a steer on a western farm, but acres, many or few, will represent much money—five dollars to one hundred dollars an acre—invested in the bare land, and that will mean a costlier steer.

Sheep prices are now high—very high. There is a big demand for both mutton and wool, and the supply does not keep pace with it. Hog prices have averaged very high for a period covering the past four years. Now, the handling of hogs is almost as big a part of the packers' business as is the killing of cattle—and the sheep department is not much behind either of these branches. If all the packers, or any of them, were in a combine to depress prices, why should they neglect sheep and hogs? To do this would be to fall far short of the business shrewdness with which their enemies credit them.

In the last few years raisers of sheep and hogs have universally made great profits, while the cattlemen have suffered to a considerable extent—and all because of natural conditions wholly beyond the control of the packers. The cattle business has been in a transitional condition—the subject of fundamental changes which have thrown immense numbers of cattle on the market and forced a period of low prices. This is the whole story.

Sober-minded men never would cry "monopoly" in relation to the business of the packers if they understood that business. While the meat industry is probably the largest in the country, it is less known than many of far less consequence. Its character and magnitude have never been comprehensively presented to the public. A comprehensive and authoritative compendium of this industry is to be found nowhere outside of government reports. Unfortunately for a clear understanding of many questions of public moment, these reports are not popular and widely-read literature compared with the "best sellers" of the book-stores. The last census figures are now five years old, of course—the facts they express are still "new" to a majority of the people—but because they are the only official figures extant they will be used to illustrate points in this article. The conditions they reflect still hold in the main. Any changes that might be made in them would strengthen rather than weaken the packers' case.

This alleged "Beef Trust" cannot be seriously regarded as monopolizing the dressed-beef and packing industry in the face of official government figures, backed by the investigation recently made under Mr. Garfield, Chief of the Bureau of Corporations of the Department of Commerce. It is admitted now, even by most of its critics, that the so-called "Beef Trust" handles less than fifty per cent. of the beef and packing industry of the country. This is the statement in Mr. Garfield's report, and, as will be shown later, that the industry holds this moiety of the business only by the advantages of foresight, superior organization, and superior business methods. But these advantages are not enough to give such a "combine" a monopoly of the dressed-beef and meat-packing industry.

The industry is too widely distributed, is too deeply rooted in too many widely separated localities to be monopolized. Without having analyzed the figures in this express relation, I feel safe in asserting that the packing industry holds a higher relative position as to value of product in *more states and cities* than any other industry in the country.

There were, in 1900, nine hundred and twenty-one meat-packing establishments in the United States. This figure did not include establishments that slaughtered only: those were classified separately in the census reports of 1900; it included only those that both slaughtered and performed the other functions classed under the head of packing and utilization of by-products, and there were nine hundred and twenty-one of them.

These packing-houses were distributed among forty-two of the forty-nine states and territories and the District of Columbia. All of these states except nine had three or more packing-houses. In twenty-six of them the industry amounted, back in 1900, to more than one million dollars each annually.

This industry ranked first among manufacturing industries in value of product in each of six states—Illinois, Indiana, Iowa, Kansas, Missouri, and Nebraska. In California it ranked second only to sugar-refining. Maryland—never thought of as a meat-packing state—gave it fourth place, as did Minnesota, Oregon, and Washington. It held seventh place or higher—probably much higher now—in widely different states, each having several specialties—New York, Ohio, Kentucky, Rhode Island, Texas, and Wisconsin. Does that look as if it were a monopoly confined to the five or six Western States where the large packers had their business concentrated?

In looking at the packing industry alone we find, of course—Chicago having developed it—that Illinois led in 1900 in volume of product, with over one-third of the total. Kansas and Nebraska were second and third, on account of the large houses at Kansas City, Kansas, and South Omaha, Nebraska, each with almost ten per cent. of the total. But what other state do you think ranked fourth in packing-house products? None other than New York, and Indiana was fifth. The next eight named in the order of their rank as packing states were: Missouri, Massachusetts, Iowa, Pennsylvania, Ohio, California, New Jersey, and Wisconsin. Each of those eight produced in 1900 one and seven-tenths to five and one-half per cent. of the country's total packing-industry product.

The packing industry is thus widely distributed and deeply rooted because it is a *supply-and-demand business* that can be established wherever natural supply and demand conditions permit. No other industry of comparable importance is so close to the people in all aspects or so closely knitted into the fabric of national prosperity. It has grown as the nation has grown, and has contributed its share—I might truthfully say more than its proportionate share—to general and individual prosperity. In the fifty years from 1850 to 1900 the total paid for the raw material used in the industry, the product of American farms, was raised from nine million, four hundred and fifty-one thousand and ninety-six dollars to six hundred and eighty-three million, five hundred and eighty-three thousand, five hundred and seventy-seven dollars—was multiplied by seventy-five.

The period of most rapid development was the decade from 1870 to 1880; that was due to the development of the refrigerator-car for shipping fresh beef, development of the export trade, development of the canning feature of the business, and development of by-product utilization—one hundred and twenty distinct by-products now being made by the packing-houses. During the next decade, 1880 to 1890, the great Chicago packing-houses took commanding position.

The stock-grower or cattleman who makes a business of raising, finishing, and marketing beef-cattle does not need to be told that there is no combine of packers to depress the price of his stock. His smaller neighbor, who raises a few cattle as a "side line" in connection with his farming operations, and who does not closely follow market reports, crop reports, and who does not analyze conditions in the cattle business, may be easily led into error along with the unanalytical general public.

Go where you will throughout the cattle country—the grazing states or the feeding states where range cattle are "finished" for market—and you will find cattlemen among the prosperous men of the community. They are a wide-awake, energetic, progressive class. They are raising and marketing cattle, not for fun or for their health, but to make money.

They do make money—not in all seasons of each year, nor every year; few do in any business; but in the long run a good proportion of them have prospered. It is obvious that any combine of cattle-buyers that would put down prices so that these men could not make money would ruin the cattle business, cut off the live-stock supply upon which the packing industry depends, and thus ruin the packing business. Do business men knowingly commit business suicide?

The packers could not by a combination control the market price of live cattle. The very nature and character of the business forbids that. They would not do it if they could. That would be business suicide. The packers have greatly helped to develop the cattle business. That they have been moved by no philanthropic motive, but by cold business sense, does not alter the fact. They had a business of their own, they saw opportunities, wide as the world, for developing and extending that business; but to utilize those opportunities they had to encourage development of the cattle business. This they did, not merely by building an industry that furnished an every-day market for cattle, but they did it by personally lending their aid to specific developments of the stock-raising business.

The most casual review of the period prior to the development of beef-refrigeration and beef-canning shows that weak, uncertain market conditions were the rule. The entire trade was on a hand-to-mouth basis. The cattle-raiser had no stable market, and speculators fleeced him at every turn.

CAR LINES AND THE PEOPLE

In the first twenty years after refrigerating and canning were commenced by the Chicago packers, the population of the country increased fifty per cent.; *the number of cattle shipped and marketed by the cattlemen of western ranges and middle western corn-farms increased 500 per cent.* The marketing of this immense production *at a profit* was made possible only by the packers' energy, the utilization of by-products, the establishment of new markets, and by pushing the business to the utmost bounds of the earth.

There is no attempt to deny that at certain times the prices for live cattle are unprofitably low—the natural result of a temporary oversupply; but it should be remembered that the price which the packer receives for his beef invariably reflects this depression of the price of his raw material. He has no control over the one price or the other; both are governed by supply and demand.

CHAPTER VIII

PUBLIC PREJUDICE INEVITABLE

THE public prejudice against the packer is more than popular—it is almost universal. In my opinion, this prejudice is inevitable, and will always continue *without regard to the manner in which the packing business is conducted.*

This is because the packer deals in a food product of universal and extensive use—a food which furnishes the *basis* of living instead of being a small and only a comparatively incidental part thereof. In the very nature of things the prices of meats are bound to rise rather than lower; the contraction of the range, as I have already explained, is sure to continue and also sure to increase the cost of producing beef; the price of corn-lands is steadily and inevitably rising, and this, also, means the increased cost of raising and feeding meat animals of all kinds.

Of course, the consumer does not naturally take all this into consideration when he goes into the retail market to buy his meat; he only recalls that the price he is paying is higher than he used to pay, and so he damns the packer and lets it go at that. It is a perfectly safe thing to do.

I do not remember a time when the retail prices of meats were satisfactory to the consumer, and I do not believe any one else does. The cost always seemed too great to the consumer—even when it represented, as it sometimes has—a direct loss to the packer.

Again, we shall miss something essential to the understanding of the packers' secure position in the disregard of the public if we fail to take account of one point of human nature: the fact that there is, in every person who goes to market, a sense of rebellion against the fact that certain things *must be bought*—and bought practically every day. Compulsion in anything is not pleasant, and there is no joy in buying something that *has to be bought* to prevent the pangs of hunger. Consequently there is, universally, a feeling of resentment against the necessity of such expenditures.

Those purchases which give pleasure are not the basic necessities of life; they are the luxuries, or at least the finer comforts. And it is human nature to think how many of these coveted things could be bought with the money which *must be paid* out for meat and the other articles of food. Thus the daily meat bill seems to stand constantly between the consumer and some coveted comfort, some article of beauty, some greatly desired luxury or pleasure. And because it does so stand—so far as the feelings of the purchaser are concerned—it provokes an unreasoning resentment of which the packer is invariably the convenient target.

All of this, to my mind, is something inherent in the situation which exists independent of the manner in which the packers conduct their business, and will exist without regard to how they may conduct it in the future. And this feeling is *always there to be appealed to by the agitator.* It makes persons of the fairest intent and of the best training and environment the ready victims of violent shocking prejudice against the packers.

And one of the most pitiable features of the matter is that the individuals who thus yield themselves to this prejudice are wholly sincere. Their attitude is impersonal; they have no personal relations with or knowledge of any of the packers; they are simply made receptive, by the general trait of human nature of which I have spoken, to the falsehoods and misrepresentations put out by the yellow magazines and the public speakers who live on this kind of sensationalism.

The bitterness and the venom of their feelings against the packers is not generally realized. One way in which it is brought directly to the packer is by means of denunciatory letters from persons who are strangers to the packers receiving the epistles. From the mass of these I select one as representative of the extremes to which the cunning and long-continued campaign of prejudice-building against the packers will move a man from whom one might naturally expect fairness and consideration.

This letter is from a minister of the Gospel, the pastor of the First Presbyterian Church in a thriving and prosperous city of Michigan. This minister of the Gospel is wholly a stranger to me and I do not believe that we have ever come in contact with each other, no matter how remotely. There is, therefore, no personal reason for his rancor. But here is the letter:

"I am writing to inquire whether it pays, in your judgment, to come into the possession of millions by the methods of the sneak and the wrecker? To say nothing of the faring awaiting such a robber in the world to come, it seems to me the contempt and bitter execrations of millions of one's fellow men can scarcely be atoned for by the possession of great wealth.

"When Marshall Field died the other day the entire country mourned. No one denied his right to the millions he had amassed honorably. But were you to pass away tomorrow, the news would be received with general satisfaction from one end of the country to the other. No business besides your own would suspend operations for an hour. It should be a bitter thing for you to realize the loathing and detestation in which you are held in every place where

your unfair, small-souled, cruel methods are becoming known.

"It is the hope and prayer of a great many that the courts will send you behind the bars for some of the crimes with which you are charged. Degenerates can feel the stigma conferred by the penitentiary who are insensible to the blights of moral condemnation. But consider Depew, Herrick, Odell, Durham, and Cox, who are now suffering from the recoil of the public conscience. I give you a text—See Matthew 23:29-36."

Is not this a most unnatural letter from a minister of the Christian religion to write a stranger? Would you sit down and put upon paper such an expression of hatred against an avowed enemy? I think not! A man to do this without any personal provocation whatever shows that his mind has been powerfully wrought upon—persistently and systematically warped through a cunning appeal to his prejudice—and this is generally cloaked under the disguise of an appeal to conscience.

I assert my profound belief that a letter of this kind from such a source would be absolutely an impossibility without such a campaign of persistent misrepresentation and organized villification as that to which the packers have for years been subjected.

Think of the occupant of a Christian pulpit, in this enlightened day, going out of his way to write a stranger a letter of such studied venom, rankling with a hatred that would have done justice to a barbaric Highland clansman, in the old days, expressing his contempt of a feudal enemy! And yet I have no doubt that, personally, this Christian pastor is a kindly man—perhaps a very gentle and amiable one—who really loves justice and delights in doing good.

To me he is simply an example of the absurd extremes to which this propaganda of slander against the packers has warped the judgment and the sentiment of thousands of the best people in the country.

Because, as I have said, a compulsory expenditure for a hard-and-fast necessity naturally creates a kind of latent sentiment, the slanders of certain magazines and periodicals have found an acceptance otherwise impossible; they have persisted and increased because it was found that the people read these charges eagerly—and that they therefore made circulation.

These publications have shrewdly and skillfully cultivated the impression—so easy to cultivate, for the reason I have explained—that the packers are the natural and inevitable enemies of the people, preying upon them as the wolf preys upon the flock. Only, fully to delineate the character of the packer as it is depicted by the sensational magazine, the wolf should be a double-headed monster with one set of jaws busy hamstringing the cattleman, while the other is closed upon the throat of the consumer.

Letters of the kind I have cited are the legitimate fruit of such a sowing as the sensational publications have indulged in; that actual violence does not follow is no fault of the most radical of these publications, who are not looking for the facts from which to lead their readers to a fair conclusion.

Do you think that this conclusion is not warranted? Then let me assure you that I have received, through the United States mails, a typewritten letter without place or date, addressed to myself and reading as follows:

"We have been reading about you and your kind of commercial tyrants in McClure's. We would think that you tyrants and —— —— ought to get tired from your continuous gorging upon the sweat and blood of the people of the nation. What

are you and —— —— aiming at, anyway? What are your ultimate objects? Have you not got enough of this world's goods already? Are you pirates seeking to enslave the people? Why not commence to be men with human hearts and try to be fair and just?

"We feel that the authorities at Washington are going to clip your cruel claws very shortly, but we also feel that personal punishment and a little terror on the Russian plan will in a large measure aid in accomplishing effectively the work in hand.

"We have, therefore, organized ourselves into a VIGILANCE COMMITTEE for the PUBLIC SAFETY, and propose to use dynamite and assassination to help suppressing you commercial vultures of the nation.

"We propose to be fair and just in our operations, and all accused tyrants will be justly tried before our tribunal, and if found guilty and sentenced to death, the con-

demned will be duly notified of the penalty they are to pay, and which punishment cannot be escaped; if not to-day then to-morrow. This is the plan which has been so successfully in operation in the execution of the Russian political tyrants by the Committee of Public Safety there.

"We beg to notify you that you have been tried by our tribunal, and your death decreed. PREPARE FOR THE INEVITABLE! You may temporarily avoid the execution of this sentence, but your time will shortly come. Our officer who has been appointed to execute this sentence has already been appointed, and you may be assured that he is prepared to sacrifice his life in such a worthy cause. The game is worth it.

"THE VIGILANCE COMMITTEE FOR THE PUBLIC SAFETY."

This letter indicates upon its face that it has been brought out by the magazine attacks upon the packers. It speaks for itself in that particular. The only comment I care to offer upon it is that it is the natural and legitimate outcome of the wanton agitation to which the packing industry and the private-car-line industry have been subjected.

Another popular method of inciting prejudice against the packers is to represent them as throttling the railroads of the country and forcing unfairly low freight rates on dressed meats and packing-house products. Some railroad officials have encouraged this misrepresentation. At banquets and elsewhere they have tossed off jauntily-worded expressions of rate-making that sound well in the ears of the inexpert and strengthen the belief that "the packers make their own rates."

Mr. A. B. Stickney, president of the Chicago Great Western Railway, has frequently indulged in word-pictures of the poor trembling railroads in the clutches of the voracious packers. One of his most often-quoted statements runs as follows:

"In fixing the rate on dressed meat we" (the railroads I presume) "don't have very much to say. The packer generally makes the rate. He comes to you and always makes you feel that he is your friend. Then he asks how much you charge for a certain shipment of dressed meats. The published tariff may be twenty-three cents a hundred, but he will not pay that. You say to him: 'I'll carry your meat for eighteen cents.' He says: 'Oh, no, you won't; I won't pay that!' Then you say: 'Well, what will you pay for it?' He then replies: 'I can get it hauled for sixteen cents.' So you haul it for sixteen cents."

That sounds convincing; to the average reader or listener it appears to be a freshly written, undried page out of the every-day experience of a railroad manager, and sounds as if it might be a verbatim report of what took place only the day before in Mr. Stickney's office. Now, what are the facts?

The rate on dressed meats from Missouri River points—Kansas City, Omaha, St. Joseph, and Sioux City—to Chicago is twenty cents a hundred, and eighteen and a half cents on through business. That was the rate when Mr. Stickney made the statement quoted above. It had been the rate for more than three years. It will continue to be the rate for more than three years longer.

That rate was fixed by a formal legal contract between Mr. Stickney's road, the Chicago Great Western, and the *Missouri River packers*—a contract *executed in the summer of 1902 and made binding for seven years. And Mr. Stickney was so well pleased with that contract that he made execution of it the occasion of a circular letter to the Great Western stockholders, in which he explained* (and almost boasted of) *what a good bargain the railroad had driven with the packers*—a bargain that meant an advance of fifteen per cent. over rates previously prevailing, and that meant a fifty per cent. increase in the railroad's net earnings.

Mr. Stickney also made it appear in his circular that the new rate was as high a rate, the Great Western believed, as could be justly exacted from the packers in view of the small margin of profit in the packers' business. But let Mr. Stickney speak for himself on rates. Here are extracts from that "private and confidential" circular which, as I have said, Mr. Stickney sent to Great Western stockholders under date of August 4, 1902, the italics being mine:

"It gives the management pleasure to be able to announce that the company has executed identical contracts with each of the packing companies doing business at Kansas City, St. Joseph, Omaha, and Sioux City, by which the packers agree to route over the Chicago Great Western lines at least a certain percentage of the entire output of their plants, at definite rates, for the term of seven years. *The rates are a substantial advance over the rates which have heretofore prevailed.*

"The aggregate revenue which these contracts secure to the Chicago Great Western Railway on the present volume of business is estimated to be fourteen million dollars, and if the business increases as rapidly in the next seven years as in the past, approximately twenty million dollars.

"These contracts cannot be understood without a knowledge of the magnitude of the packing industry. The published report of Swift & Co. gives the amount of its sales last year at the enormous sum of two hundred and twenty million dollars. Presumably its chief competitor, Armour & Co., did substantially as much, and it is probably safe to estimate that the aggregate sales of the other packers amount to enough to make the grand total fully seven hundred million dollars. More than half of the aggregate business is the output of the plants at the Missouri River cities mentioned, and is effected by the contracts.

"The narrowness of the margin of profits is even more surprising than the magnitude of the transactions. The report of Swift & Co. (the only report available) gives the information that on sales during last year, practically in a retail way, aggregating over two hundred and twenty million dollars of perishable commodities requiring the greatest care to guard against serious loss, the entire profits were only about three million dollars, or less than one and one-half per cent.

"With such a narrow margin of profit it is easy to see that freight rates are an important factor in the packing-house business. It is estimated that the annual freight bills amount to three times the annual profits. Hence, any increase in the rates which have been established for years, to which the whole business has been adjusted, is a serious matter.

"And the packers being *willing to agree to a permanent, substantial advance of more than fifteen per cent.*, the management felt that its duty to the stockholders demanded that it should accept the opportunity to secure, for a term of years, this substantial advance in rates. Accordingly it has entered into identical and lawful contracts with each and every packer doing business at Kansas City, St. Joseph, Omaha, and Sioux City.

"In consideration of this contract on the part of the railroad, the packers agree to ship over the Chicago Great Western lines in each and every month during the full term of seven years at least a certain percentage of the entire output of their respective packing-houses, and of all such packing-houses as they, their successors and assigns, may hereafter own or control, *and to pay therefor the full published tariff rates,* regardless of any lower rates which may be offered by other railway companies.

"These contracts, unlike the 'maximum rate contracts,' are legal, and therefore enforceable in the courts.

"There is no way by which the other lines can reduce the rate or quantity.

"It is certainly satisfactory to know that so large a volume of gross revenue, approximately one million dollars per annum, is already secured by contract for the Omaha and Sioux City lines, now under construction, as soon as they are completed."

It is instructive to notice the difference between Mr. Stickney talking to the general public about freight rates and President Stickney telling his stockholders in a confidential circular what a good bargain he has made with the packers. I have only to add that loose-tongued talk and undigested generalizing by men supposed to speak authoritatively has had much to do with prejudicing the public mind against the packers.

One of the most flagrant of all the many misrepresentations which have recently appeared in certain magazines is this statement: "Beef is hung up in the refrigerator-cars. There is a space beneath on the floor of the car. It has been charged that this space is sometimes full of dressed poultry, eggs, and so on. Poultry and eggs take a high freight rate; but, thus packed, *Armour gets them carried for nothing.* . . . How much of such business goes on no one knows, *but it has been shown to exist in numerous cases.*"

Nothing could be falser than this statement. It is untrue in every respect and particular. The older men in the employ of Armour & Co. are witnesses to the fact that my father's instructions were most strict on this point; that he guarded against anything of the sort by pointing out both the dishonesty and the foolishness of any practice of that nature.

The same precautions against the possibility of that sort of thing on the part of an over-zealous employee have been taken by myself. There is not a man in the shipping department of Armour & Co. who does not thoroughly understand that an attempt at such a practice would bring him instant dismissal.

Any person believing that such a thing would be done by any packer can quickly rid his mind of such a notion by going to the freight department of any railroad handling packing-house business. The roads' inspectors are not only on the platforms from which the cars are loaded, but they have access to the books of account and to the very invoices from which the collections are made from the persons or houses to whom the cars are shipped.

The statement is not only utterly false, but it is absurd, and any freight man who knows the actual processes of shipping from a packing-house will say so. They know that a thing of this kind could not be done without detection, and that an attempt to do it would be silly and suicidal.

CHAPTER IX

ANOTHER CONTRIBUTION TO PROGRESS

WHENEVER the people of this country—or any of them—come to cast up their score with the packers, there are some things which cannot in justice be overlooked, although they are so commonplace as to be accepted as a mere matter of course.

The packers' contribution to economic progress in the way of by-product utilization has been enormous. It is impossible to realize the extent to which this touches almost every physical element in our everyday life—from making worn-out and barren lands yield bounteous crops to supplying the buttons on our coats. If all these by-products were suddenly stricken from commerce the void would astound the world, and the result would be everywhere considered a dire public calamity.

"WASTE NOT" is the packer's creed, and his scientific faithfulness to it—inspired by self-interest—is actually one of the most fruitful sources of economic advantage to the people of the civilized world thus far brought about by the aid of the laboratory of the scientist.

Let us see just how much this by-product utilization means to the grower and the consumer of food-animals as well as the people in general.

In the old times packing was done in the winter. The first change in method was the use of ice and the commencement of summer packing. This started in hog-packing, but with the introduction of the refrigerator-car beef was killed largely in the summer. About this time some of the packers adopted the method of packing and shipping meats in tin cans. The refrigerator-car permitted beef to be killed near where it was grown, as it was cheaper to pay freight on five hundred and fifty pounds of carcass beef than on one thousand pounds of live animal.

Immediately following this the railroads endeavored to advance the freight on dressed beef so that they could still continue to ship the animals alive on the hoof, as they were afraid that their tonnage would be materially reduced. It was soon demonstrated that under the new system their beef tonnage was greater and their old live-animal tonnage smaller, and though the dressed-beef rates east were much greater per pound than for live animals, yet the freight on five hundred and fifty pounds of dressed beef is less than on one thousand pounds of live animal.

The four hundred and fifty pounds of non-edible material was largely thrown away, although the hide and tallow were utilized. Later, some of the waste product was used in the manufacture of glue.

Nitrogen being the chief element in plant food, and this being abundant in the great mass of refuse matter originally thrown away as hopeless waste from all the packer's processes, a most important economic advance was made in the step which turned this large volume of scrappage into fertilizer.

It is good sense and for the best interest of the world that all material not needed to feed, clothe, and heal the world should be returned to the ground as food for plants, to grow more grain, to feed more cattle, and to feed more people. Thus is the circle completed by the packer.

All the cunning of the chemist has been called into service to *save,* to make the most of every scrap of material in hand, and to discover new ways in which some element of waste may be diverted from uselessness to use.

Hundreds of valuable products are now made and shipped all over the world from materials which, under the old methods, had little or no value. Thousands of people are employed in manufacturing these products. The technical schools are constantly being called upon for young men to aid in solving new problems in by-product utilization. New plants are being built requiring material, machinery, and labor in their construction. Success in by-product utilization in the packing industry has directed the attention of other industries to this important element in industrial administration.

All this directly affects the people and has been of great benefit to them. The investigator in medicinal and other lines is constantly calling on the packer for material to aid him in his work. In the pharmaceutical line much has been done of benefit, and many ills are helped by pharmaceutical preparations of animal origin.

In the fertilizer line many sections are given over to growing products which could not be profitably grown without the use of fertilizers. The upland cotton section of the South has been *made* by the use of fertilizer in the growing of cotton.

Sandy soils in sections climatically favorable have been developed into large truck-farming districts through the use of fertilizers, as the soil, without fertilizer, is practically sterile. In the manufacture of fertilizer the packer has done his share in saving material formerly permitted to go to waste.

The importance of fertilizer to the agricultural and fruit-growing interests of the country calls for a word of detailed explanation regarding this interesting and valuable by-product. Fertilizer is food for plants the same as corn for cattle and meat for men. In the early '80s, Peruvian guano being very high and a demand having arisen for commercial fertilizers, it was suggested to the Chicago slaughterers that the meaty and bony materials deposited in the bottom of the large tanks used in rendering lard and tallow be dried and used as a source of nitrogen and phosphorus in compounding fertilizers, mixing with them acid phosphate made by dissolving rock phosphate with acid. This demand came first from the East, and was the beginning of the fertilizer industry in Chicago.

A hot air dryer was designed for this work. This led on to utilization of still other materials that formerly were largely wasted. The blood accumulating in the killing of cattle was boiled to coagulate the albumen, and was dried and also used as fertilizer. The water in the rendering tanks also is evaporated and the solid portions dried and made into concentrated tankage.

These materials, tankage and blood, and ground waste hoof and horn scraps, are valuable chiefly for their nitrogenous contents. Blood and hoof meal, containing almost as much nitrogen as nitrate of soda or sulphate of ammonia and being very available although not soluble in water, possessed certain chemical advantages over nitrate of soda and sulphate of ammonia when compounded with other fertilizer ingredients. These materials were for a number of years sold to the manufacturers of complete fertilizers, but in the early '90s my father began the manufacture of complete fertilizers himself, combining with his material potash from Germany and phosphates from Tennessee, Florida, and South Carolina.

This end of the packing business has grown very rapidly, and Armour & Company now have plants at many points in the United States for the manufacture of fertilizers, shipping their blood, bone, and tankage to those manufacturing points, making their sulphuric acid and acid phosphate at these various plants, and supplying the trade of the immediate section. Fertilizers are made of different analyses for

Bird's-eye view Armour and Company's Plant, Union Stock Yards, Chicago, Illinois

different soils, climates and crops—a quick acting fertilizer being prepared for truckers who must get their product into the very early markets and a slower acting fertilizer for winter wheat. Practically all of the cotton in the United States, with the exception of a part of that grown in Texas, is grown with fertilizers, and nearly all of the winter wheat producing states are now largely using fertilizers.

In connection with the fertilizer works is prepared blood albumen which, up to six years ago, had been imported entirely from Europe. It was claimed it could not be made in this country; the climate would not permit, etc. This product is used by the calico printers in fixing certain pigment colors, such as ultramarine, to the cloth on which they are printed. Blood albumen is similar to that in the white of an egg, and when the temperature is raised to a certain point it coagulates. A paste ink is made by mixing liquid albumen with the pigment, and this ink is used in printing the cloths. After these are printed, they are run into a room and steamed at a temperature approximating two hundred degrees. This action coagulates the albumen in the ink, rendering it insoluble in water, and mechanically fixes the color in the cloth.

In addition to this, albumen is used by the tanner in finishing leathers; by the makers of certain extracts, and by sugar manufacturers for the clarifying of liquors.

Dried blood is used as a fertilizer, and is also beginning to be used quite largely as a stock food. It is the most concentrated food we have, containing something like eighty-seven per cent. of protein, and gives exceptionally good results when fed to horses and other live stock in connection with their regular feed. The Kansas City experiment station has discovered that a disease known as "scours," which greatly troubles young calves, can be cured by the addition of a little of this dried blood in the skim milk on which they are fed. This discovery is saving thousands of calves every year. Certain portions of tankage are ground up into meat meal and largely sold for the growing of poultry and for the fattening of hogs and live stock. Bone is coarsely granulated and also used for the same purpose.

The furniture of the country is glued with the packers' glue. A great deal of the wool used in clothing is made from sheep slaughtered by the packers. One of the largest sources of curled hair is the switch from the tails of cattle. Bristles are cured for the brushmaker. A large portion of the soap manufactured comes from the tallows and greases prepared by the packers.

The first steps in by-product utilization cannot fail to interest those who like to trace the lines of economic progress.

In the early days of the packing industry, as I have already remarked, very little use was made of the offal and refuse other than the hide, tallow and grease. The blood, cattle feet, the head and other refuse was buried or hauled out on the prairies and buried. The horns finally became of some value, and after the horn pith was removed they were shipped to Europe, and there manufactured into horn buttons, combs and various ornaments; but no manufacturing of this sort was attempted in the United States until within recent years.

About the first use made of the offal at Chicago was in the latter part of the '70s, when a small glue manufacturing plant was established. There was no attempt to buy material from the packers, but, as the story goes, the cattle heads and feet buried by the packers during the day were dug up at night, hauled to the glue factory and converted into glue and bone fertilizer. Soon after this the packers realized that the material had some value, and began to charge for it. In the meantime, quite an industry had been built up in the manufacture of glue, and in the year 1885 Armour & Company purchased the Wahl Brothers' glue factory. This was the first entrance of the packers into the manufacture of glues.

The industry has grown rapidly, Armour & Company having the largest glue plant in the world, which not only uses the raw material furnished by its own packing houses, but also buys largely from other packers. They manufacture all grades of glue—bone glues, hide glues, etc.

The phosphate of lime in bone is held together by nitrogenous binding material from which bone glue is made, and which can be taken out of bone by cooking at a low temperature, or the lime can be leached out of the bone stock by acids. After the glue is extracted, the liquor is concentrated in vacuum, chilled by refrigeration, cut up into thin layers and dried on wire screening in large hot air dryers. Then it is sold as sheet glue, broken glue or ground glue, as the trade may demand. During the process of glue boiling, the tallow and grease is extracted and skimmed off and the bone and meaty matter deposited is dried, ground up and sold as fertilizer.

Gelatin is also made at the glue works from selected "calves' stock." The process is one requiring great care and skill.

Brewers' isinglass is made from animal tissues and is used by the brewers in the clarifying of their liquors. This product is a transparent white article sold in sheets and resembles the finest grades of gelatin in appearance.

In the bone-cutting department, a branch of the fertilizer works, are made white knife handles and handles for knives, razors, etc., shaped, carved and dyed to resemble stag horn. Here are also made collar buttons, pipe mouth-pieces, bone screws used to connect the pipe mouth-pieces and the bowl of the pipe, dice, little square blocks or "dummy teeth" used by dental students in practice work; nursing-bottle rings and shields, and bone buttons of all kinds.

In the manufacture of fine steel parts of guns, bicycles, automobiles, and other pieces of fine mechanism, it is necessary to case-harden the surface of certain working parts; that is, steel is fashioned into its final shape when soft and these parts are then put in a large iron box and surrounded by hard granulated bone. This box is then fastened together, luted and put in a furnace. There it is heated to a certain color and the contents dumped into water. The result is that the steel has been case-hardened. In other words, a very high carbon steel has been formed on the outside of the parts thus treated, making them extremely hard. Bone is also used to make the blue color on rifle barrels, etc.

A word regarding the utilization of hoofs: These are assorted into three grades; the white hoofs being used in the manufacture of a certain grade of buttons which closely resemble the better quality of pearl buttons. The striped hoofs are flattened into plates by pressure under heat, and are used in the manufacture of hairpins; they are also used in the manufacture of buttons. The black hoofs and hoof scraps are used in the manufacture of cyanide and chrome. They are also ground up into a fine powder and used as a nitrogenous fertilizer for the growing of grapes and other special crops.

Glycerin is another important by-product. At a certain time in the boiling of soap salt is introduced into the kettles, which carries to the bottom the glycerin of the fat together with the lye that has not been used in the saponifying of the soap. This spent lye is drawn off and from it is manufactured glycerin. In the manufacture of soaps all grades are made, from the very finest toilet and shaving soaps to the mottled German scrubbing soaps. Washing powders are also prepared from soap stock.

Packing-house laboratory products, the results of original research by scientists of the first class, are employed every day by physicians, surgeons, dentists, and chemists throughout the world. More than thirty recognized therapeutic agents of animal origin are produced in Armour & Co.'s laboratory. Among them are the pepsin and pancreatin which physicians use in treating digestive disorders.

There is a product of thyroid glands that is employed in treating cretinism or idiocy. Another is suprarenalin, used in the most delicate surgical operations to stop the flow of blood. To illustrate how closely the by-product feature of the business is gleaned, the suprarenal glands of more than one hundred thousand sheep are required to produce one pound of suprarenalin, and when produced this suprarenalin is worth more than five thousand dollars a pound.

The oil extracted from the wool of slaughtered sheep is used in various emulsions and toilet preparations, being peculiarly soft and soothing to the skin. Anhydrous ammonia is another by-product of the packing-house laboratory and a very extensive business is done in this article.

Extract of beef is a product which serves to illustrate the progress of the packing industry during comparatively recent years, and the rapid evolution from the slaughter-house to the wonderful packing plants of to-day, where every department is conducted under the most advanced scientific methods.

The process of making beef extract is simple. Realizing that South American and Australian beef extracts, which are made from wild cattle usually killed for their hides and bones, could not possibly possess the delicious flavor of beef extract made from the carefully fattened domestic cattle of the United States, the American methods have been directed towards the preservation of the true beef flavor in a concentrated form. Tons of fresh meats are used daily for making beef extract. After the extract has been taken from the meat, the remaining fibre is dried, ground or powdered and sold largely in foreign markets, where it is used in various forms of animal food.

The liquors cooked from the meat are concentrated in vacuum pans to either solid or fluid consistency, as may be desired. It is then a pure, concentrated extract of beef. Gouffe says that beef broth is the soul of domestic cookery, but how to get that "soul" from the old-time recipe, "Take a shin of beef," has proved a difficult problem to many housekeepers. Now, with a jar of American beef extract at hand, the cook has, without time or trouble, the best beef broth for making soups, sauces, beef teas, etc.

As the consumption of beef extract has

grown, so, too, have its forms of preparation. To-day one may obtain it in the solid or fluid form, or in combination with asparagus, tomatoes and other delicious vegetables.

A few years ago a physician friend of Mr. Philip D. Armour asked him why he did not make a nutritious preparation of beef that would be a food upon which life could be sustained independent of all other foods. He said this was needed particularly for the invalid and the convalescent, and pointed out that his means and position as a packer made such an expensive investigation possible. Mr. Armour gave instructions to his chemist to start the work, and as a result "Soluble Beef" was placed upon the market.

"Soluble Beef" differs from beef extract in that it is the real substance of the meat—fibre and all—predigested and concentrated. An idea of its nutritive value may be had by comparing it with raw beef juice, which

Putting up Meat Extracts, Armour and Company's Plant, Chicago, Illinois

has enjoyed a high reputation as a food for the sick. Analysis of soluble beef shows that it contains fifty-four per cent. food value; raw beef juice contains from two to five per cent. On this basis one teaspoonful of soluble beef is equal to from ten to twenty-seven teaspoonfuls of beef juice. It may be used with hot water without affecting the nutritive value. The advantage of such food to the invalid or convalescent is evident to all.

It is impossible to pass the matter of the packer's products without reference to the really important subject of oleomargarine. During the Franco-Prussian war, Mege Mouries, forced by the conditions existing in Paris, ascertained that the oil expressed from beef suet was similar to the oil in butter fat, and when churned in milk made a very good substitute for butter, and the resultant product, oleomargarine, was given to the world. This product does not contain as large a percentage of butyric and other volatile acids which give rancidity to butter as does butter itself. Therefore, oleomargarine keeps better than butter. To the lumberman, the miner, the sailor, this product gives a reliable supply of butter food which he could not obtain if he were dependent upon butter. The animal oils are churned in milk, worked, salted and handled precisely as butter and the product is butter made by chemical methods rather than by Nature. The attitude of the dairy distributing interests of the country, as voiced by Congressional action, has, for the time being, greatly injured the oleomargarine business, but it is, nevertheless, an article of high merit and great economic importance.

It seems impossible to believe that the laboring portion of the community is to be deprived of a healthful and cheap article of food at the instigation of the manufacturers of a competing article, *simply because of its competition.* The internal revenue laws controlling the sale of oleomargarine before the passage of the Grout bill protected the consumer in that they required original packages of oleomargarine to be branded, and a stamp tax of two cents a pound to be attached to each package. All of this was under the control of the Internal Revenue Department, the same as is the manufacture of cigars and other taxable commodities. It was sold under GOVERNMENT CONTROL. There was no possibility of the manufacturer selling "oleo" as butter without incurring very great risks and penalties, and it is safe to say that no manufacturer attempted it.

Under the Grout bill, however, the restrictions are so severe that the packer is practically compelled to look to foreign countries almost exclusively for a market for this clean, wholesome and eminently serviceable product. One of the unjust features of the present regulations is that which prohibits the manufacturer of oleomargarine from using any coloring; in other words, from giving it the attractive appearance of butter. As it can be colored just as wholesomely as butter, the restriction is purely an arbitrary handicap imposed solely to limit its use—for people accustomed to use butter will not readily and extensively use a substitute which looks unlike it. When this product of the packing house is given the commercial chance to which its merits entitle it, the results will be of great advantage to the poorer people of this country and to the cattle raisers.

Commissioner of Corporations Garfield's report on the beef industry shows that the legislation against oleomargarine in 1902 caused a very marked decrease in the value of the fat derived from beef cattle. The average net value of fat from cattle killed in Chicago, his report shows, fell from $4.31 a head in the second half of 1902 to $2.65 a head in the second half of 1903. For all packing centres combined, the average fall in value between these two periods was $1.29 a head. Such decrease in the value of products derived from cattle decreased, of course, the price that could be paid for cattle.

The packer who could make the most out of these products could afford to pay and did pay more for the live steer than his competitor who was not so progressive, and in consequence he got his pick of the cattle. The stock-grower was benefited by the higher price paid for the live animal, and the people were benefited by the lower selling cost of the beef over the old method.

One cannot eat his cake and have it, too, and the reward that the packer received was that of increased business and the decreased cost per head of killing cattle, owing to his much heavier kill; but, in order to cash in that reward, he had to give both the stock-grower and the public a part of the benefit of by-product utilization.

CHAPTER X

ONCE MORE THE PRIVATE-CAR LINE

THE interests of the private-car line and the packing industries are so intimately connected that the future of either must necessarily involve, to a considerable extent, the future of the other.

Broadly speaking, they are equally the objects of attack on the part of mistaken or malicious agitators; but the sharpest fight seems to be focused on the private-car lines, and, therefore, I shall place emphasis on that more acute line of campaign.

If the hostile legislation now aimed at either or both of these industries becomes law, there is no question that the men who are pushing it will have the satisfaction of having dealt a hard and perhaps fatal blow to the packers and to two great industries which have done more, I believe, than any other two industries to give the whole people the cardinal comforts of good living: wholesome fresh meats and fresh fruits and vegetables.

But their satisfaction will not end with crippling these agencies of administration to the common needs of humanity. They will also awake to the fact that theirs will be the credit of dealing a staggering blow to scores of other industries—to the cattle-raising business, to the calling of the fruit and the vegetable growers, to the entire agriculture of the country, and, finally, to the scores of other industries which are sensitively and inseparably interrelated with the animal, fruit, and vegetable industries.

Believing this, it seems to me that whatever threatens the future normal and legitimate development of the business in which the private-car lines and the packers are engaged directly concerns not only every business man of this country but every individual in the United States.

Therefore, I offer no apology for meeting the charges, attacks, and criticisms which have been made public in certain magazines, periodicals, and newspapers, and in the public utterances of the enemies of the private-car lines and of the packers.

Although the assaults have, in many cases, been personal—and bitterly so—and the provocation is strong to deal with them in kind and to show their personal animus (as I can in every instance), they will be met in general terms, but, I believe, with sufficient definiteness.

In some points it will be necessary to refer briefly to matters already touched upon, but only for the purpose of giving a clear and adequate bird's-eye view of the war that is being waged to disorganize industries which are indispensable to the American people, and to make their future a record of struggle, and perhaps failure, under the handicap and hardship of unfair and unwarranted legislation.

The misrepresentation employed by the anti-car-line champions cannot overcome the force of this simple fact: The men who pay "extortionate rates" to the private-car lines—that is, the actual growers and shippers of fruit—*never have voiced a serious complaint against the car lines, and do not now favor the anti-car-line agitation.* This does not mean there are not some individual or association complaints.

I feel perfectly safe in saying that more than ninety per cent. of the growers, where private refrigerator-cars are operated, are in favor of keeping the private-car lines with their refrigeration service and their exclusive contracts in operation as they are. The better business man the fruit-grower is, and the more experience he has had with commercial enterprises other than fruit-growing, the more heartily does he speaks out for the private-car line and its exclusive contract.

Such men understand the risks in any business; therefore they appreciate what an advantage it is to have a responsible concern bound to furnish, at the instant needed, *good* cars, *clean* cars, and *enough* cars, with *prompt* and *certain* icing—advantages which they seldom had when railroads attempted to furnish the refrigeration, or when several refrigerator-car companies competed for their business—advantages which they cannot hope to get for years to come, if they are deprived of private cars and forced to depend on railroad refrigeration.

Men of this class appreciate, too, that *good* service must be paid for, and that *poor service is dear at any price;* and *service* in the handling of perishable berries and fruits *is the first consideration,* as every successful grower will testify.

If the most extravagant and misleading

comparisons that have been made between private-car and railroad refrigeration rates were true, the difference would amount, on peaches, say, to ten to fifteen dollars a car. The practical and successful grower reasons thus:

"When I get good refrigeration and reliable service for ten or fifteen or even twenty-five dollars a car more than I would pay for poor refrigeration service, the extra money is well invested. That extra ten or fifteen dollars a car will mean, in almost every instance, from fifty to one hundred dollars a car added to my net returns by reason of my fruit getting to market in good condition."

Is it not a distinct credit to furnish a quality of refrigeration service for which growers are willing to pay an advanced price? I believe it is. Here and there, of course, one finds a "kicker." Absolute unanimity would be impossible as human nature is now constituted, but the kicker's grievance can always be traced to some individual and usually accidental happening.

Illustrating this general point as to the quality of service, let me quote from one of several letters now on my desk from Koshkonong, Missouri, the largest peach-shipping point in Missouri—letters from growers who had heard that the private-car-line service might be withdrawn from the 'Frisco Railroad System this year. Mr. T. M. Culver, who manages five hundred acres of Elberta peach orchard, writes:

"I, as well as a lot of other peach-growers at this place, have planted and are still planting thousands of trees, and not by any means the least incentive of our large plantings is the excellent service we get from the Armour Car Lines. If you say that we will have no more Armour cars in which to ship our peaches, it will be the greatest disappointment to me I have ever met with in my peach-growing experience.

I firmly believe it would be a good many years before such a service as Armour's could be had on the 'Frisco.

"I have shipped hundreds of cars of peaches, and not one complaint have I ever had as to condition on arrival, nor have my commission men ever complained to me about extortionate charges. Neither have I any complaint to make along this line." And Mr. Culver closes with the assertion that the loss of the private-car service "will be disastrous to peach-growers."

I could quote similar expressions by scores from fruit-growers—practical and successful fruit-growers—in all parts of this country from Michigan to Georgia and from Delaware to California.

Now just a final word on the real animus of this fight on the private-car lines. Fruit-handling commission men—not all commission men, but some—have been forced to come out in the open and admit that they are trying to kill the private-car lines if they can, and cripple them if extermination is not possible. Hostile legislation is their machine gun.

One of the chief arguments they have used in enlisting the aid of representatives, senators, writers, and other men who make sentiment and legislation is the cry that the private-car lines are monopolistic in character, grasping in purpose, and that they seek to control the handling and refrigeration of fruits.

Portraits of the leaders of this band of public-spirited commission men adorn the pages of magazines and periodicals in which appear the arguments inspired by them. They are hailed as the champions of anti-monopoly, the protectors of the people in general and of the fruit-growers in particular. Their mission is to get the oppressed grower out from under the heel of the private-car-line magnate; to liberate him from the control of the monopolist.

Please keep all this in mind while you

read a few extracts taken from the address of President Streight, delivered before the Annual Convention of the Western Fruit Jobbers' Association, held in Omaha on December 28, 1904. My quotations are made from the pamphlet sent out by the Association as the official report of its proceedings. The italics are my own. Here is the significant declaration made by the official head of the Western Fruit Jobbers' Association:

"The great percentage of the commodities we handle are extremely perishable. On this account our business is of a more hazardous nature than any ordinary merchandising. The perishable nature of our commodities alone is sufficiently hazardous without the aid of *fluctuating prices,* overstocks, *unfriendly,* illegitimate, and unprofitable *competition.*

"We should have an organization which would take in every legitimate and honorable jobber of fruit and produce in the Middle and Western States, with the object of *reducing the hazardous nature of our business* to the greatest possible extent.

"We cannot change the commodities we handle, but we can collect and disseminate information for the benefit of each member. *We can form local and district* organizations and eliminate to some extent UNFRIENDLY AND UNPROFITABLE COMPETITION, lessen the overstocking and fluctuating of local markets, and become *business and social friends* instead of simply *unfriendly competitors* throughout the territory."

Another member of the Association also addressed the Omaha meeting in the following significant language:

"But conditions are changed. A large part of the business is now done on the f. o. b. plan, or else, if consigned, the shipments are confined to a few reliable houses who have the absolute confidence of the shipper, and the shipper feels that by loyally standing by the receiver through thick and thin he will achieve greater results than in the 'old rainbow-chasing days.'

"One of the greatest drawbacks of the present method of buying f. o. b., especially has it so proved the past season, *is the disposition on the part of our representatives to outbid one another.* I have a case in mind that occurred in Louisiana last spring. Two representatives of Minneapolis houses, actuated by their jealousy, wanted all the strawberries from a well-known point. The result was that prices advanced from three dollars and twenty-five cents one day to four dollars and twenty-five cents the next day, *this without increasing the production one iota.*

"Another case is the Van Buren deal. Texas is practically through shipping when Van Buren begins, and, as the latter is the first in Arkansas to move, their berries are eagerly sought after.

"Last season there were probably ten men for every car the first week, and, of course, the *local shipper took the advantage of our necessities, boosting prices out of sight, cleaning up from three hundred dollars to five hundred dollars a car, while by a* LITTLE CONCERTED ACTION ON OUR PART just as many berries could have been secured *at probably a dollar a case less.*

"This is a point we should seriously consider, not that I propose a combination on prices so much as I do to avoid bunching our men at one particular point and bulling the market. Moreover, *a close relationship* should be encouraged among our buyers and solicitors. Let it be understood that when we cannot land a shipment ourselves, or else have all we can use, we see that a member of the Western Fruit Jobbers' Association is favored.

"Now let me put the question right here: Will every member of this Association make it a point to instruct his field-man, as well as do it himself, to use his influence to advance the interests of our Association among shippers, and, when it is impossible to secure a shipment for his own house, make an effort to see that some other house in our Association is favored in preference to an outside concern?"

Could any appeal for combination against the grower and shipper be more clear than this convention declaration? I think not. The most ingenuous and unsophisticated fruit-grower cannot fail to understand the hostile intent of language like this.

And again I ask: Could there be a clearer, a more definite and authoritative verification of my statement that the work of the private-car lines *in bringing competitive buyers into the fruit-growing districts, there to hustle for business and bid against each other,* has been of inestimable benefit to the grower and shipper?

It is an official confirmation of my statement made out of the mouth of an avowed enemy of the private-car lines.

It is a confession, published in the camp of the enemy, that competition on the part of the buyers—*forced into the field by the private refrigerator-car*—results in great gain to the grower, brings him far higher prices, and has liberated him from the domination of the commission man.

There is not a grower in any district served by the private refrigerator-car—at least none who was there in the fruit business before the coming of the car—who will not admit that the *buyers followed the car into the field,* and that the private car revolutionized the commission business, bringing the commission man to the grower, whereas the grower had before been obliged to seek the commission man and accept his terms.

The real attitude of the commission men toward the growers and toward the question of "combination" for the purpose of squelching "unfriendly competition" can only be fully appreciated in these utterances of the official heads of the Western Fruit Jobbers' Association. The cry of "monopoly" and "combination" certainly comes with rare grace from this organization!

Some of these commission men lose all sense of perspective—and of humor, too—when they undertake to explain how they love high-class refrigeration for its own sake, but are nobly battling against the monopoly.

One spokesman, in a recent deliverance, grows eloquent with virtuous rage when he contemplates the "horrible conditions" imposed upon refrigerator-car service by the "Armour monopoly." Further along he avers that Armour cars "are to the refrigerator-cars of the whole country *but as a drop in the bucket.*"

If "but a drop in the bucket," I ask, How can it be a monopoly that is strangling the fruit industry? Again, the spokesman of this particular commission-man coterie makes the specious plea that he and his associates are not trying to drive private-car lines out of the business, but are only seeking to "regulate" the rates; then to prove his words he quotes a lawyer-like statement to the effect that "legislation cannot be framed," under the Constitution, to "prevent formation of independent car lines for hire of cars to railways."

Quite so, but these very men who are so sure of what cannot be done under the Constitution are working night and day at Washington for a law that will restrict refrigeration service charges to the actual cost of ice, pound by pound. Such a law would put the refrigerator-car lines out of business as effectually as would a law flatly prohibiting them.

Efficient *refrigeration service,* with adequate car-supply, ice-supply, icing, re-icing, and inspection, cannot be performed by any line, railroad or any other agency, for actual cost of ice, unless it is done at an actual loss.

But with all their evasions, these commission men cannot get away from this fact: They desire to drive out of business the private-car lines that furnish efficient refrigeration service. For assistance in this they rely much upon that trait in human nature which always enables a falsehood to travel faster than the truth, and they have chosen an apt time for such a campaign—a time when the public mind has been poisoned by "yellow" agitation against everything bearing the name of corporation, and by demagogic appeal for political effect.

All this was admitted—inadvertently, no doubt—when the president of the National League of Commission Merchants, in convention at Milwaukee last December, summed up the results of the League's anti-private-car-line fight in these words:

"But we feel that it is at *this session of Congress* that our labors must be rewarded and the necessary legislation enacted, and *we feel encouraged to believe that, if not from the merits of our cause, then* FROM THE POLITICAL EXPEDIENCY *of our situation, this will be brought about.*"

Could a confession be plainer?

But I hope that I shall not be misunderstood as classing all commission men with the ones I have been discussing. Nothing could be more remote from my thought. Many of the strongest houses in the trade have no sympathy with the fight being made in the name of the Western Fruit Jobbers' Association and the National League of Commission Merchants. Members of these organizations have not hesitated to take a stand against them.

A case in point is J. D. Hendrickson, of Philadelphia, a former president of the National League, who went to Washington last year and testified that private-car-line service was a necessity in the perishable fruit business. Mr. Hendrickson is both commission merchant and peach-grower; and I believe practically every commission merchant like him, who also knows the fruit business *as a grower,* stands with him on the side of the private-car lines.

No longer ago than January 20, 1906, F. Newhall & Sons, of Chicago, members of the National League, wrote to the *Fruit Trade Journal and Produce Record,* the official organ of the commission trade, protesting against the anti-car-line movement. I quote a few characteristic sentences:

"We believe our firm is only one among a great many that have been benefited, instead of injured, by the private-car lines.

"Our experience with refrigerator-cars before the private-car line came into existence was a sad one. You could seldom get a refrigerator-car from the railroad company when you ordered one, and often when you did it was a very poor one, not suitable for the purpose that you wanted it for. . . . We have made fifty claims for loss and damage on fruit shipped in refrigerator-cars furnished by railroads to one for fruit shipped in private-car lines, such as the Armour Car Lines.

"While they (the refrigerator rates of the private-car lines) are higher than charged by some railroads, it has always seemed cheaper in the end, because our goods arrived in better condition and were worth more to us than the difference in refrigeration charges.

"The private-car line refrigerators have been a very great benefit to us in furnishing good refrigerators to move our shipments in throughout the winter season, when no charge is made for icing. . . .

"If we find icing charges too high, let us go to the heads of companies controlling the lines making the unreasonable charges and try to induce them to readjust their rates. We can *accomplish more in this way than by trying to put them out of business. . . .*

"We believe in a square deal for all, and we know there is a very large element in the fruit and produce business in the United States which believes as we do."

The business genius and commercial pioneering, the enterprising, organizing and executive abilities of the original packers have been among the most potent influences in building up the country. Why shouldn't they be entitled to some credit for it?

They were looking out for themselves when they were building those businesses. Of course they were. But that is no answer. If we examine the intimate personal annals of the heroes of history, we shall find but few who started upon their respective roads lighted only by the pure white flame of holy resolve to uplift humanity. Blink at it as we may, there is a touch of self-interest in the best of men.

The Revolutionary fathers, as we reverently call them, sought, in the beginning, only to relieve themselves from distasteful governmental restrictions and odious taxes. They *grew* into opportunity and resolve to build a new nation on a new foundation. And to the end of their labors—but it detracts not at all from the honor due them—they were upborne by the desire to better the material, worldly condition of *themselves* and fellows. Lincoln, running for the legislature while he was still a primitive country storekeeper and an unsuccessful one; Grant, volunteering for the war of the rebellion, were hardly moved entirely by visions of liberating a race or saving a nation. They were simply breaking into those fields where they knew *their* talents would have freest play and *where they believed they could do the best for themselves.* So true an idealist as the late John Hay said in his article on "Franklin in France," in a recent magazine: "It is the infallible sign of decadence in a man or a government when he undertakes work which cannot pay expenses."

The original packers were born into a commercial age. They were filled with the spirit of the times. They saw opportunities lying at their feet—opportunities to widen their own commercial field and that of all around them—opportunities to create new enterprises—opportunities, if you please, to make two blades of grass grow where but one grew before. Made as they were, they could no more smother their energies than the born artist can keep his fingers out of the paint box. They jumped in, and the pioneering, the creating, and the organizing achieved by them and men of their stamp, has played a large part in making this country the marvel of history in rapid development and commercial expansion.

Look again at the much-belabored car

lines that are but one branch of the business my father developed—a by-product, as it were, of his packing-house ventures—and note their influence in just one state—Georgia. That state is enjoying a season of prosperity unknown since the war. Higher cotton prices have brought a measure of that prosperity, but the fruitful peach tree must be credited with a share. And the peach tree brings profit to the planter only by means of the refrigerator-car.

That statement is not too strong by a hair's weight when you consider Georgia. The peach cannot go far to market in quantities unless it goes under ice. Georgia has no nearby markets of consequence. Its people could not profitably grow peaches for market on a large scale if they could not get facilities for shipping under ice. They will all tell you so. And they were encouraged to grow peaches for the larger markets, in the first instance, because they were offered such facilities by the refrigerator-car service of the private-car lines.

As a typical instance of what has been done by the private-car lines for peach growers and for the community at large in half a dozen counties of northwestern Georgia, I have in mind a little settlement in Chattooga County. It is tucked away "miles from nowhere" under the shadow of the mountains. For transportation, it depends upon a little railroad that is less than one hundred miles long and that has no road but itself to depend upon—no alliance with any large system. It is, or was, a country of discouraging clay hills and rock outcrops, scrubby timber and uninviting aspect.

The first shipment of peaches out of that community was nine years ago last summer. Summer before last—the crop last summer was light—that one locality shipped about four hundred car-loads, and those car-loads brought back hundreds of thousands of dollars, net. The little town that forms the chief shipping-point is putting on airs with brick buildings. It has a prosperous bank, whose stockholders boast that it paid thirteen per cent. the first year of its existence. The entire community is prosperous. Men who were living in cabin-like "shacks" a few years ago are building handsome residences and sending their children away to school. Men who were poor a few years back—poor even for north Georgia "crackers"—are now well-to-do—wealthy as wealth counts among farmers—and talk glibly of $10,000 and $12,000 cleared in a season on fifty or sixty acres of peach orchards. Land that was worth no more than $1.00 an acre ten years ago is now held at $8.00 and $10.00 an acre, even without peach trees on it; when bearing an orchard it brings three times that amount.

Fruit and vegetable growing in Georgia was truly an infant industry, a dwarf, in fact, until the private refrigerator-car lines gave it a chance to grow. Small quantities of peaches reached an outside market by express in boxes to Savannah and then to New York and Philadelphia by boat. But there was no life in the business. It was cramped by restricted and expensive shipping facilities. It was not worth the name of an industry. The product was scarcely taken into account by the railroads as an element in the freight producing resources of the state.

In the early '90s the private-car lines went into Georgia and began to show its people how they could reach the fruit markets of the entire country by means of the refrigerator-car service, and land fruits at destination in condition to command a good price. Three or four seasons of experiment with private-car service were so conclusive to the growers and so stimulated the growing of peaches that the railroads were simply overwhelmed by the demand upon them for moving the peach crop. They had to call on the private-car lines for refrigerator-cars, and then revise schedules, rearrange plans, and make special arrangements for moving the fruit cars. Since that time fruit and vegetable growing has climbed by leaps and bounds to the rank of a great industry, and is still advancing to a higher position.

Unfortunately there are no accurate data showing how the industry has grown from year to year or showing exactly the point it has reached. Some of the railroads traversing the fruit belt have undertaken to tabulate the peach business, but their figures are incomplete. They are sufficient, however, to warrant the statement that fruit and vegetable growing ranks second only to cotton in bringing prosperity to Georgia people, and in earning for Georgia her favorite title of "Empire State of the South."

These incomplete railroad statistics show that peach growing alone is a considerable industry in more than sixty different counties in Georgia. The state peach-growers' association estimates seventeen million peach trees in their state. Not all are in bearing, but are very near it. In north Georgia, peach trees are planted one hundred and sixty to the acre; in south Georgia as high as one hundred and ninety-six to the acre. The total acreage in the state, therefore, must be in the neighborhood of one hundred thousand acres; and it is growing all the time.

This immense peach acreage is divided up among individual growers—farmers—whose orchards range all the way from five acres to twelve hundred acres. A peach tree in bearing will yield anywhere from one crate to five crates of peaches, according to its location, its condition and the care given to it. This being so, it is apparent that the peach growing industry, fostered by the private refrigerator-car, is worth literally millions of dollars annually to the state of Georgia.

This money is not for a few; does not go into the treasury of a "trust." It is distributed direct among thousands of individual growers. To the great majority of peach growers it is the largest part of their annual income; and to a considerable degree it is *new* income, *additional* income. If it were not received from peaches it would not be received from any product. For the point cannot be over-emphasized that much of the Georgia peach crop is grown on land that was all but worthless and unproductive until planted with peach trees.

But the benefit to the people from the development of this industry does not stop at the peach grower. Think of the additional millions of money put into circulation in each district by the expenditure for labor in the orchard, for the harvesting and handling by laborers and railroad crews, for baskets and crates, for fertilizer, and so on. The fertilizer industry, already a large one in the South, has been stimulated by the new fruit and vegetable growing until the railroads are hardly able to handle its product during the three months in which it must be handled. The effect of all this can hardly be appreciated unless studied at close range. The state is enjoying unparalleled prosperity and it is *prosperity of the people throughout the country districts* like the prosperity we heard of from the West a few years back—the farmer building a new house, presenting a piano to his daughter and setting up his son with a young horse and a top buggy.

Higher priced cotton has been a big element, but the peach is entitled to a share of the credit. Proof of it? In almost every little town of north Georgia you can see the proof—brick and stone buildings built out of peach orchard profits—handsome new residences built by men who have made a fortune in peaches and have moved into town from the hills to take life easy and give their children advantages.

Cantaloupe culture is coming to be an important part of this Georgia development, thanks to the private car. Statistics on cantaloupes are even less complete than on peaches, but our agents report a rapid growth in the industry. Many farmers in south Georgia are devoting some of their very best land to growing cantaloupes. It requires no stretch of imagination to see cantaloupe growing lead to ventures into other fruit and early vegetable crops for which there is always a demand in northern markets when they can be landed there in good condition and at reasonable expense—as they can be by the means of the private refrigerator-car.

There is a reflex benefit from this diversification of the southern farmers' crop. For years cotton was king, and king-like, lorded so absolutely as to become at times disagreeable. Over-production of cotton sent prices down to the no-profit point. Recent effort has been to restrict cotton production—even to the extent of letting cotton land lie idle and so better the price. This has had some effect. Every acre of cotton land put into fruit and vegetable growing reduces by so much the possible cotton acreage, and helps, by so much restricted production, to boost cotton prices without the wastefulness of land lying idle.

The packing industry has helped teach the western and northwestern farmer to diversify his product by feeding cattle for beef instead of blindly depending upon wheat only, or corn only, and suffering in years of one-crop failure or over-production. The private-car-line refrigerator service is helping teach the southern farmer to diversify by raising the fruits and vegetables that northern markets want and cannot get elsewhere in sufficient quantities.

THE PACKERS, THE PRIVATE

CHAPTER XI

JUGGLING THE FACTS

NOT long ago the press of the whole country united in tribute to the worthiness of a great merchant who lay dead.

The immense commercial fabric which Marshall Field left behind him was a monument to the practice of one fixed business rule: "To sell *always* a *good* article at a fair price." The private-car-line interests which I represent ask to be judged by no more lenient standard.

We have been in the refrigeration-service business for twenty years. We have been successful—so successful the demand has required that every dollar earned by the car lines, *and more,* has had to go back into new and improved equipment and organization to take care of the new business that was made possible by the service. I do not believe this could have happened if the car lines had not sold *a good article at a fair price.*

An article in a recent magazine contains this paragraph: "By giving his business to one railroad, and taking it from another, he (Armour) could almost make or ruin the profits of the companies concerned. Here was the Père Marquette Railroad, for example, over which was shipped the bulk of the fruit grown in the rich districts of western Michigan. The Père Marquette was a weak railroad, hungry for more traffic. Armour went to the eager officers of the Père Marquette and guaranteed to give them forty cars of meat a week, in return for which the Père Marquette agreed to use none but Armour's fruit-cars for its fruit-shipments. The Père Marquette had a few refrigerator-cars of its own, with which it served it customers; but under the new contract it could not supply its own cars to the people along its own lines."

That statement is misleading as a whole and specifically false in detail. The writer of it never would have ventured to make it if he had sought the facts. He could have easily learned them from the Interstate Commerce Commission's records, which are open and accessible to all. Now what are the facts? They are simply as follows:

The Père Marquette Railroad undertook, in 1901, to furnish refrigeration for the fruit business along its line, using a few old refrigerator-cars of its own and borrowing from other lines. That season, with its bad service and consequent heavy fruit-losses, is still remembered with a shudder by Michigan peach-growers.

For the following season, 1902, the Père Marquette Railroad made an experimental contract—an exclusive contract—with the Armour Car Lines. When that shipping season was over, the Père Marquette Railroad officials made inquiry among the growers and shippers as to how they had been served and how they liked the exclusive arrangement. The response was so heartily in favor of the Armour service and the guarantees of the exclusive contract that the Père Marquette immediately made another exclusive contract with the Armour Car Lines for three seasons.

The first, or experimental, contract was executed July 18, 1902; the second was executed on December 23, 1902.

The meat-carrying contract between the Père Marquette Railroad and Armour & Co. was executed on August 1, 1904. *Therefore, this meat contract, by means of which the writer in question says the fruit contract was extorted from the Père Marquette, was not made until more than two years after the execution of the first fruit-car contract and nineteen months after the second fruit-car contract, and there was nothing unusual or "special"* in the meat-carrying contract. Other packers made similar agreements with the Père Marquette at that time—in the summer of 1904.

And right here it should be said that this gross misstatement comes as near the truth as any of the charges brought to show how Armour, as a packer, sandbags the railroads into giving his car lines exclusive contracts for the refrigeration of fresh fruits and vegetables.

Now there is a sequel to the Père Marquette case; and the sequel should be interesting to those whose sympathies have been wrought upon by sensational writers until they mourn the fate of Michigan peach-growers in the clutches of "Armour and his monopoly." It is the more interesting because all orators against the private-car lines in magazines and in Interstate Commerce hearings must have made the Père Marquette situation their pet "horrible example."

The private-car lines' exclusive contract for fruit-refrigeration service on the Père

Marquette Railroad expired in 1905, about the time the road went into the hands of a receiver. That contract was renewed within thirty days by the receiver representing the United States Court. It is an *exclusive* contract, too.

The receiver of the Père Marquette is Hon. Judson Harmon, of Cincinnati, a lawyer of national reputation, who was Attorney-General of the United States in President Cleveland's Cabinet. The court behind him is a United States Court, superior in its powers even to the Interstate Commerce Commission.

It will hardly be alleged, I think, that Judson Harmon, acting as an agent of the United States Court, has made this contract as a part of a trade for a haul of "forty cars of meat a week." Nor is this emphatic endorsement by the growers and shippers of that line due, as our enemies may contend, to the fact that the Père Marquette road is in the hands of a receiver.

The fact is, these growers and shippers have stood by us loyally from the first year of our operations in Michigan, and will doubtless do so as long as we continue, as we have in the past, to give them value received.

It seems hardly necessary to say that Judge Harmon, as receiver, made this contract only after he was thoroughly satisfied that it would be the best for the railroad and best for the growers and shippers along its line. His method of satisfying himself is especially interesting in view of the widely circulated statements to the effect that the fruit-grower has been robbed by the private-car lines and is struggling to escape from its clutches.

The question of getting ready to move next summer's fruit crop came up to Receiver Harmon in the natural course of railroad administration.

To determine exactly what ought to be done about it—and perhaps moved somewhat by current criticism of private-car lines and their exclusive contract system, especially as to the Père Marquette Railroad—Judge Harmon ordered a house-to-house canvass, as it were, of the territory concerned. The officials of the road went out along the line, met the growers and shippers at the respective shipping-points, invited frank and outspoken expressions of their experience with the refrigeration service furnished by railroads as well as by private-car lines, and asked what they wished for the coming season.

The stenographic report of those conferences shows that a preference for the continuation of the previous arrangement—for private refrigerator-cars operated under the exclusive contract that is alleged to be "throttling the fruit industry"—has been and is practically unanimous from one end of the railway to the other. The report abounds in expressions like these:

"The rate is a secondary consideration; quality of service comes first."

"I prefer the Armour car, at double the price, to any other car we ever had."

"Deliver us from refrigerator-cars borrowed from other railroads! We want cars that we know will come to us clean and in good condition."

"If the Père Marquette were sending in their cars free and furnishing ice and all, I would prefer the Armour cars and pay the present rate."

The difference between private-car line refrigeration rates in Michigan in 1900, or before, and those subsequent to that year has been the subject of persistent and willful misrepresentation. This juggling of the truth has been the more dishonest because it has sought to justify itself by emphasizing a technicality.

The private-car lines' Michigan tariff for 1902 *was* higher than for previous years.

The reason for it was this: Prior to 1902 the Michigan railroads *paid for all the ice* used for the initial icing of all fruit-cars before being loaded, for re-icing after being loaded, and for re-icing *en route* to the East, the heaviest part of the business; the shipper paid for no ice except that used for re-icing cars *en route* to the South and West.

The car lines' tariff, therefore, entirely excluded the cost of the *initial icing and re-icing eastward; it covered only the service* eastward and only service and re-icing to the South and West. After the first contract was made the railroads stopped furnishing ice, and the ar lines' tariff then had to be made high enough to cover the new expense (namely, the *cost of all ice* both for initial icing and re-icing) as well as the service before. This change in rates brought no additional profit to the car lines.

Free icing at the railroad's expense was never practiced, so far as I know, in connection with the fruit business anywhere but in Michigan. It was discontinued there —after the 1900 contract—for the simple reason that the railroads could not afford it. The traffic-manager of the Père Marquette, Mr. A. Patriarche, in sworn testimony before the Interstate Commerce Commission, in June, 1904, made clear the fact that free ice ate up fifty per cent. of the revenue from fruit-shipments. The car lines' part in the change was explicitly stated to the United States Senate's Committee on Interstate Commerce, May 15, 1905, by Mr. George B. Robbins, president of the Armour Car Lines, as follows:

"In 1900 we furnished refrigeration to shippers of Michigan peaches, and under the railroad rule or classification then in effect, *the railroad paid us* for, or absorbed, the cost of ice both at loading-stations and *en route,* and our rates were based on these conditions. I cannot recall a similar rule having ever been in effect elsewhere.

"In 1901 or 1902 the Michigan roads changed this rule and discontinued furnishing the ice free, and we advanced our charges to cover the additional cost of ice to us.

"We had nothing whatever to do with this change in rule, and our profit was not increased by the increased charge for the refrigeration over the previous abnormally low one. The matter is one entirely between the road and the shippers and is not chargeable to the car lines in any way."

By cunningly exploiting the rate-advance, which was due entirely to the railroads' discontinuance of free ice, as has been shown, the car lines' critics seek to smother this further fact.

Michigan peach-growers have prospered under the good service furnished by the exclusive contract and full refrigeration-tariff as they never prospered before; they have said so; and, as we have seen from

Receiver Harmon's inquiry, the peach-growers along the Père Marquette heartily favor a continuance of the service.

It is pertinent to the rate question to reiterate this statement: In practically every section where the Armour cars operate under exclusive contracts, refrigeration-rates are lower than they were before such contracts were made, and as fast as economies can be effected in organization and management, and as business increases, rates are further reduced. This policy has been consistently pursued and will continue to be.

This shows the way in which, from start to finish, the magazine agitators have made their case by unsupported personal statements—cunningly selected half-truths and imaginings dressed up to look like facts, and all trimmed to fit preconceived theory. It is clearly impossible, for reasons of space, to consider all their charges.

It is worth while, however, to touch

briefly upon a few other facts that have been persistently—and adroitly—garbled by these writers.

Throughout the studied efforts to "make a case" against the private-car lines, facts have been juggled or wantonly suppressed to give these impressions:

That the only expense attached to refrigeration service is the cost of ice; that the car lines secure exclusive contracts with railroads by some secret, underhand method, or by coercing weak railroads, and then arbitrarily "charge what they like."

That the largest, best-equipped, and best-operated railways furnish refrigeration service "of the highest character" at lower rates than the car lines do.

That the refrigeration-service charge is "extortion," as applied to the fruit and vegetable grower, and a "burden" upon the "perishable food-supply of the nation."

Let us deal specifically with these misrepresentations.

Private-car-line refrigeration service, I repeat, is not an ice business; it is a *highly specialized service*—a typical product of this specializing age. In the professions—law, medicine, engineering—the specialist commands the highest price for his work. That business house—in manufacturing or in merchandising—is most successful whose department heads have best learned how to specialize.

Why apply a different standard to refrigeration service? There was a time in the railroad business when passengers and freight were carried in the same train. The freight-train of to-day is better, faster, and safer than the mixed passenger and freight train of former days; the "regular" train is still better; the "limited" is better yet, and a ride on it costs more.

Specializing—nothing but specializing, in car-building, engine-building, track-building and operation—has produced the Chicago-to-New York eighteen-hour train.

There is as much difference between good refrigerator-cars and bad refrigerator-cars, between good refrigeration service and poor refrigeration service, as there is between the good and poor in any kind of business. As in other lines of business, too, a good refrigeration service costs more than poor service—much more—and is worth more—worth much more—than the extra cost.

It is, therefore, misleading, not to say dishonest, to assume that all refrigerator-cars and all kinds of refrigeration service are alike, to compare rates on that basis and to call the higher rate "extortionate" without taking into account the service value given for that rate. As well say that every suit of clothes is equally good and all suits ought to be sold at the same price.

Let me recapitulate, briefly, the main points of the service:

Refrigeration service of the first class—the kind private-car lines furnish—must be thoroughly organized and manned by specialists, and must be comprehensive in its field of operation, equipped to give the shipper service *in* and *to* any part of the country.

Its cars must be specially designed for carrying highly perishable berries and fruit; it must keep those cars in the best possible condition, and permit them to be used but rarely and discreetly for carrying anything but fruit, berries, or vegetables.

It must maintain, the year round, an expert and expensive force of men to prepare in detail, and long in advance, for the car-supply, ice-supply, in most districts bracing and stripping lumber, labor, etc., to meet the varying demands of each season in each district where it operates.

It must carry a still larger force to manage and supervise loading and icing stations during the shipping-season, men who receive traveling-expenses as well as salary, and who are idle at the expense of the car lines practically three months in the year.

It must maintain throughout the country icing-stations, icing-inspectors, and men to handle the ice for prompt re-icing of cars and whole trains in transit.

It must maintain car-building and car-repair shops near all the large fruit districts, and often gangs of repairers in the field, for hardly a car makes a trip without needing some repairs. The employees in the field are, in most cases, practical fruit men as well as refrigeration men, who can show the inexperienced grower how to handle his crop.

And, finally—but by no means least—the car line must assume the risk. It must buy and store great quantities of ice and organize its force of employees before the first fruit-tree blossoms. It may make all the extensive preparations for a record crop that turns out a total failure and converts preparations into a dead loss.

On the other hand, it may prepare, on the best estimates obtainable, for an average crop and be called upon to handle a phenomenal one; and if it falls short of enough cars, ice, or men to handle an unexpectedly heavy or fast-ripening crop, it must stand the consequences.

At this writing we are put to extreme measures, because of the open winter, to secure an ice-supply for next season. At one northern point alone, we are spending one hundred thousand dollars for an ice-machine to meet this emergency. Again, we are buying machine-ice where we should have natural ice, and shipping it from remote points. All this heavy expense involves no advance of refrigeration-rates. The burden falls on the car lines, not on the growers.

Without the guaranty of an exclusive contract, as I have tried to make clear, the private refrigerator-car line could not afford to assume the risks of this business;

neither could it afford to furnish the service at the price it charges. Experience has demonstrated that to the satisfaction of growers, shippers, and railroads. Experience has demonstrated, too, that the exclusive contract is as much a benefit to the car lines' customers as it is to the car lines. It is a guaranty to the grower, the shipper, and the railroad that when the fruit crop ripens in a given territory there will be an ample supply of first-class refrigerator-cars to carry the fruit in, and an organized refrigeration service to protect the fruit on its way to any and every market.

The refrigerator-car line and the exclusive contract exist solely because they *have been proven to be an economic necessity in the handling of perishable fruits and vegetables.* They owe nothing to favoritism and none is practiced by means of them, directly or indirectly, for or against any railroad, car line, grower, shipper, or receiver of shipments.

Some of the most important contracts held by the Armour Car Lines are in that great fruit belt extending through Florida, Georgia, South Carolina, and North Carolina, where the railroads could not possibly be coerced by Armour & Co.'s "great power over the railroads" as "a great shipper." The railroads there with which exclusive contracts have been made for years *are competing lines:* the Southern, the Seaboard Air Line and the Atlantic Coast Line, the Central of Georgia, the Western and Atlantic, and the Georgia Southern and Florida, and so on. How would you force the same kind of an exclusive contract from each of three competing lines by threatening to divert freight from one to another?

Another point: Armour is not a "great shipper" over any of the railroads in that territory. The meat and provision shipments that go over these roads are not heavy through shipments on a wholesale basis—for export, for redistribution, etc.—but are the only comparatively small shipments to supply local consumption in a territory whose largest city has a population of only one hundred and fifty thousand. Shipments of that character hardly give Armour "great power over the railroads."

Now let us see what there is to the complaint that the exclusive contract gives the private-car line a "monopoly" and enables it to "charge what it likes."

A refrigerator-car line does acquire, by an exclusive contract, all the refrigeration business arising during the life of the contract on the particular railroad contracted with, and acquires it for the reasons already explained; but this contract no more creates a monopoly, in the accepted meaning of the word, than does the contract under which one paper-mill, for example, supplies all the paper of certain grades used by the United States government.

The refrigerator-car line's contract, like the paper-mill's contract, is simply an agreement that certain well-defined service shall be performed during a certain period at a certain price, and in accordance with specifically described conditions. The contract is open to competition before it is executed. The specifications or conditions as to car-supply, ice-supply, re-icing, etc., are carefully drawn to meet the requirements of the shippers and the railroads. Maximum rates are named, it is specified that the rates shall be reasonable, and the car line is not left a chance to "charge what it likes." Contract conditions having been thus prescribed, the railroad naturally lets the refrigeration contract to the car line that is best able to carry out the contract conditions. Right here I may explain that in many cases, and whenever conditions warrant it, the rates charged are less than the maximum rates named in the contract. If rates should be made too high—so high as to be burdensome—the railroad, which ought to be selfishly interested in encouraging its shippers, has the remedy in its own hands. It can annul the contract for cause.

The greater the quantity of fruit grown, the greater the volume of business for refrigerator-cars. No car line could afford to make oppressive rates that would discourage and diminish fruit-growing. And if the car line were so foolish as to do that, could the railroad afford to let it be done and so rob itself of freight-revenue from fruit-shipments?

In view of the facts I have stated as to how car lines' exclusive contracts are made, the logic of your own business experience ought to make it plain that a car line cannot "charge what it likes," and, if it could, would not desire to charge unreasonable rates.

Now for a few facts on the broad statements and broader insinuations to the effect that certain strong railroads—too strong to be influenced by the power of "a great shipper"—furnish better refrigeration at lower rates than private-car lines do.

Certain railroads do operate their own refrigerator-cars, but chiefly in the dairy products and produce business. The private-car lines which are under fire do berry and fruit refrigeration chiefly, which is entirely different from the refrigeration required for dairy products and produce. Consequently, comparison of the two, as to rates or otherwise, is unfair.

Only a very few railroads furnish refrigeration on a large scale, or pretend to furnish inspection and re-icing after shipments leave their own lines. Most of the few roads which do this, as the Santa Fé and the Gould Lines, maintain a separate refrigeration service organization like that of the private-car lines and charge relatively the same rates that private-car lines do. The railroads which make a lower refrigeration-rate do this business at a loss. I believe they would so state if asked.

The Pennsylvania Railroad and the Gould Lines have been mentioned as having refused to make exclusive contracts with private-car lines, the insinuation being that they were strong enough to defy "the octopus."

The private-car lines have no formal contract with the Pennsylvania Railroad, but one of them for years has handled most of the berries and fruit refrigeration on that road, especially in Delaware, where the most of the Pennsylvania Railroad's fruit and berry traffic originates.

And that same private-car line has been asked by the railroad to take care of the berry and fruit refrigeration on the Pennsylvania for the season of 1906. Nearly all of the Pennsylvania's own refrigerator cars are used for dairy products and produce.

The Gould Lines, as railroads, do not own

any refrigerator cars; they do own the American Refrigerator Transit Company, which is a separate corporation and is operated just as the Armour and other private-car lines are. Most of these American Refrigerator Transit cars are also used for dairy products and produce.

One Gould Line, the Denver and Rio Grande, had an exclusive contract with a private-car line which expired in 1904 or in 1905; the American Refrigerator Transit cars then replaced Armour cars on the Denver and Rio Grande for handling Colorado's large and growing fruit business. A short time ago the old exclusive contract between the Armour Car Lines and the Denver and Rio Grande, a Gould line, was renewed *at the request of that railroad and of the growers and shippers* along its lines. So in 1906 private-car-line refrigerator-cars will replace Gould cars on a Gould railroad.

These facts ought to satisfy the most captious that there is no honesty in the statement or insinuation that railroads furnish better refrigeration than private-car lines do, or furnish as good service at lower rates, or that they are forced to take our service. The testimony of railroad officials who have had to give special attention to refrigeration is practically all to the contrary.

The railroads on which the great bulk of highly perishable fruit business originates —the Southern Pacific, Père Marquette, the Atlantic Coast Line, the Central of Georgia, and other southeastern lines—long since adopted the private-car line service and facilities under exclusive contracts, *to the better satisfaction of the growers.*

The consumer, too, has a vital interest in this private-car line question. Every city housewife of moderate means knows that the season during which she can have fresh fruits and vegetables on her table is months longer now than it used to be six to ten years ago.

Then only the very rich could disregard the season by disregarding expense in supplying the table; but families of small or moderate incomes had to get their "fresh" fruit, berries, and vegetables out of cans for a large part of the year.

That condition was not changed until the refrigerator-car service, developed and made efficient by the specializing of the private-car lines, opened the way for growers in all parts of the country to grow and market early fruits and vegetables at a profit.

The truth of this will appeal to any city resident, whether a member of a family or a patron of a boarding-house, hotel, or restaurant. The moderately-circumstanced in all cities had no fresh fruits and vegetables until crops ripened at near-by points.

In Chicago, they waited for strawberries until they came from Indiana and southern

Illinois, and did not revel in them until the Michigan crop ripened; New York's economical housewives waited for Delaware's crop.

The "popular price" restaurant or hotel did not pretend to carry lettuce, fresh tomatoes, and such tender vegetables in the midwinter menu; if they appeared at all in the winter, they appeared only as occasional novelties—a stroke of enterprise by the proprietor—at a fancy price.

Patrons of all such restaurants and hotels in the large cities now have lettuce, fresh tomatoes, etc., on the daily bill of fare practically the year round and at reasonable prices.

Take the cantaloupe, for example; until within a decade it could be had for only a very short period and only at a high price. To find one on the market as early as the Fourth of July was a novelty. Now Florida cantaloupes come into the market in the latter part of May and the development of cantaloupe-growing in favorable sections, from Delaware to Colorado and California, keeps them coming to market all summer at prices that make them an every-day delicacy in families which used to deem them a luxury produced only for the very rich.

In 1897 it was estimated that not more than four hundred cars of cantaloupes were grown in the whole country. The crop of 1905, after only eight years of development, was figured at nearly seven thousand carloads. A section of the Salton Desert, California, where the private-car lines have ice-houses below the sea-level, and where the temperature often rises to one hundred and twenty-five degrees, had sixty-four acres of cantaloupes four years ago; this season it will have two thousand, five hundred acres of cantaloupes.

The practical, frugal housewife of to-day does only a fraction of the "putting-up" and preserving that she did ten years ago. Why go to that trouble? She can supply her table better and for the same, or less, sum of money because she gets fresh fruits and vegetables practically all the year. I am not saying that cost of living has been reduced by this fruit and vegetable development due to refrigeration service; *but I do say that thousands and millions of persons are able, by reason of that development, to live better for the same money.*

CHAPTER XII

THEORY *VS.* CONDITION.

NOW for a word of explanation and emphasis as to private ownership of meat refrigerator-cars. There is, it seems to me, every reason why the packers should own the refrigerator-cars in which their meats are shipped and no reason why they should not. This may appear to the layman to be a very broad statement, but I believe that a knowledge of the facts involved will bring any fair-minded business man to this view of the case.

The general public supposes that, because railroads are in the business of carrying freight, all an intending shipper has to do is to ask for a freight-car and it will be eagerly and promptly brought to the shipping-point ready to receive its cargo. But men who have had experience as shippers know that this is merely a theory, not an actual condition. The facts are that the railroad is constantly trying to get along with as few freight-cars as possible and still handle its traffic; its effort is to handle a maximum volume of freight with a minimum equipment of cars.

Now what is the inevitable result of this kind of thing? A constant shortage of cars which becomes especially acute at seasons of general freight movement. Any regular shipper in any line of business using freight-cars furnished by the railroad will tell you that one of the most serious avenues through which his business is made to suffer is that of failure of the railroad to furnish him cars when they are needed. He will tell you that this is not an exceptional emergency but a chronic and discouraging problem.

If the writers who devote their talent to attacking the packers are to be believed, the railroads are so eager to please these "favored shippers" that they would go to almost any length rather than incur the displeasure of these "freight barons." But how does the car supply problem work out in actual practice as between the railroads and the packers?

Eight weeks ago from this writing the Armour Fertilizer Works made a requisition for forty freight-cars to haul one thousand tons of fertilizer material from a certain point in Tennessee. Of those forty cars just three were actually furnished in a period of eight weeks, and a large plant dependent on the material was shut down!

This is not an exceptional occurrence. On the contrary, it is a routine circumstance. Right at the same time three ocean schooners chartered by Armour & Co. for phosphate rock loading were tied up at the Tampa docks waiting for rock which could not be shipped from the mines owing to the railway's inability to furnish cars. This delay in getting freight-cars caused a demurrage charge of $340.00 a day.

How does all this apply to the problem of the private ownership of meat refrigerator-cars? Most directly and pertinently! There are some kinds of shipping where a delay is not a serious matter, although it is always annoying and expensive to the shipper.

On the other hand, there are businesses where delay in shipping is simply fatal, where any element that interferes with *regular* and practically instantaneous shipping must be eliminated from the situation at almost any cost, for the business cannot continue under that kind of handicap. Common sense will at once indicate to any reader that the packing business belongs to this class. Not only this, but it is probably the most sensitive to this element of all industries.

Fresh meats must be shipped regularly and promptly. The world demands its meats every day and to place its supply at the mercy of an unreliable supply of cars in which to ship it would at once subject the consumers as well as the packers of meats to a peril not to be countenanced. To delay the shipment of meats, when ready to ship, means deterioration and loss. Not only must there be a reliable source of supply, but the cars must be clean and in the best of repair. Fruit or dairy-cars will not do. They are not safe. Meat must be shipped with every safeguard—the public health demands it. Railroad administration is not efficient enough to guarantee this.

The only way in which the packers can possibly protect the public and themselves from the hardships incident upon delayed shipments is to have their own refrigerator-cars which are absolutely subject to their own control and which cannot be diverted to other uses. They must know that they are to have at their beck and call, every day in the year, enough cars to handle their business and to handle it without delay or *without danger of delay.*

If I have not already given sufficient support to my statement that the railroads are not to be depended upon for a prompt supply of freight-cars, it is for no lack of definite cases in point.

In the year 1904 a large southern fertilizer maker manufactured, sold, and held in hand shipping instructions for the movement of 5,000 car-loads of fertilizer which could not be shipped because the railroads failed to furnish the necessary cars in time to get the fertilizer on the ground when the land was being prepared for the cotton crop. There were 100,000 tons of this fertilizer which, in the natural course of events, would have produced about 134,000 bales of cotton, worth more than $8,000,000, not to speak of $1,600,000 worth of cotton-seed oil and meal. Rather a heavy forfeit for the failure of the railroads to furnish sufficient freight-cars!

However, I am certain that my contention will not be questioned by any shipper who has occasion, at all seasons, to make requisition upon any railroad for any considerable number of freight-cars. No matter to how high a figure his annual freight shipments may climb, he will be found among those who suffer from the undependable supply of freight-cars—and because this difficulty is inherent in the system of railroad operation as it is now practiced and as it must be practiced for many years to come. Some of the highest salaries in the railroad world are paid to men who best approximate the solution of this difficulty at a minimum of expense. They are able executive men and accomplish wonders under the adverse conditions imposed upon them, but these conditions are such that no amount of executive genius can make the supply of freight-cars from a railroad anything but erratic and undependable. So impossible is it for railroad traffic men to cope with these conditions that we invariably make it a rule to anticipate our shipments of fertilizer and other common freight by notifying the railroads thirty to sixty days in advance of our requirements; but notwithstanding our precautions, we are invariably confronted with car shortages, ranging from ten to thirty-five cars per day at different plants.

CHAPTER XIII

AS TO EXPORT BUSINESS

A VERY important side of the packing industry, and one that is often overlooked by the layman, is the export business. Without a comprehensive view of the origin, extent, and consequences of our foreign trade in meats, meat products, and meat animals, it is impossible to arrive at a clear understanding of the mutual relations between packers, live stock raisers, and the general public.

The export business in meats and meat animals has been a great factor in opening the way for that development of the live stock industry which has completely revolutionized the character of stock raising and farming throughout the West; it has reacted beneficially upon the agricultural interests of the entire country; it has played a part, too, in giving the American people, as a whole, *better meats at lower prices* than are enjoyed by any other people on earth.

It is the export business that absorbs the surplus live stock grown by American cattlemen and farmers, and *absorbs those grades of both cattle and meats for which there is but a limited demand at home.* The effect of this is twofold. It steadies and supports the home market for live stock, and it moderates the prices of those meats which the American consumer demands; because, without a foreign market for those parts of the beef carcass which the American does not take, the slaughterer would be compelled to ask a higher price for the choice cuts in order to make up for loss on the other parts. If this export business were taken away, or even reduced, the effect of it would be directly felt, not only by the large ranchers and stock raisers, but by every farmer who raises a steer, sheep, or hog for market, and by every one of the hundreds of thousands who participate by their labor in the immense live stock and meat industry.

To appreciate the national importance of this export business in meats, meat products and meat animals in dollars, we must compare it with other branches of our export trade. The total exports of the United States last year (1905)—embracing every kind of domestic product sold abroad, the products of agriculture, manufactures, mining, forests, fisheries and the many smaller miscellaneous products—reached the splendid total of $1,599,420,539. Of this enormous total, the meat and live stock industry alone furnished about fifteen per cent., or more than one-seventh.

We are accustomed to boast that "America feeds the world." The phrase suggests, to most people, I believe, our corn fields and wheat fields and milling centres. They are all great—tremendous—and we export an enormous quantity of breadstuffs; but measured in dollars and cents, breadstuff exports rank away below meat exports. Where we sell abroad $43,000,000 worth of meat animals and over $190,000,000 of meats and meat products, we sell only $154,-000,000 worth of breadstuffs. We are great copper producers, yet we export of copper and manufactures of copper only $85,000,-000 a year. We have the richest iron mines of the world and the greatest iron and steel mills, but our exports of iron and steel and the manufactures thereof fall more than $90,000,000 short of our meat exports—$143,000,000 of iron and steel as against nearly $234,000,000 of live stock and meats.

The export of meat animals, aggregating $42,974,945, was divided as follows:

KIND.	HEAD.	VALUE.
Cattle	571,153	$41,007,375
Hogs	82,849	811,918
Sheep	191,030	1,155,642
		$42,974,935

The exports of meats and meat products used as food—including, of course, butter, cheese and milk, for which a market has been made in connection with meats—aggregate $190,660,703, divided as follows:

	Pounds.	Dollars.
Fresh beef	254,360,198	$23,246,792
Canned beef	75,208,036	7,423,071
Beef, salted, pickled, etc.	73,984,544	4,268,773
Tallow	81,702,816	3,893,986
Bacon	297,815,453	28,236,990
Hams	207,244,526	21,358,567
Pork, fresh, salt, pickled, canned, etc.	161,716,505	13,287,057
Lard	701,679,162	54,881,748
Lard compounds and substitutes	66,955,736	3,951,712
Mutton	577,636	52,238
Oleo oil and oleomargarine	192,262,668	15,503,967
Sausage and sausage meats	6,964,139	756,857
Sausage casings		2,631,193
Poultry and game		791,930
All other canned meats		1,817,786
All other meats		2,791,962
Butter	16,194,483	2,876,628
Cheese	8,229,756	935,934
Milk		1,953,512

Finishing and Testing Cans, Armour and Company's Plant, Chicago, Illinois

Total meats and provisions ..	$190,660,703
Total meat animals	42,974,935
Grand total of meat animals and animal food products	$233,635,638

American packers have done practically all of the work that has gone into making the foreign market for this immense total. They have had to do the pioneering, the promoting, the fostering and the hard industrial fighting. It has been uphill fighting, too, in the fact of prejudice, jealousy, self interest, and hostile legislation both at home and abroad. Business interests and the political interests that cater to them have been banded together in more than one foreign country to put up the bars against American meats. Similar opposition, with less reason for its existence, has been encountered at home. But the American packers have gone on steadily winning the way for American products in the face of prejudice and unreasonable trade restrictions. They were actuated by self interest, of course; but that does not detract from the value of the thing accomplished. Intelligent self interest directed along industrial and commercial channels—the impulse to get on in the world—have made this country what it is.

The period of unwarranted and unintelligent criticism of the packers through which we are now passing directly affects this very large export element in our national prosperity. Every time an American sensation-monger, political agitator or writer for the "yellow" periodicals seeks to attract an audience by attacking the packers, he aims a blow at the entire live stock and meat producing industry of the United States—at every cattleman, hog raiser and sheep grower, and every person engaged in the meat producing and meat distributing business. Every such utterance gives the foreigner just that much more ammunition with which to fight off American exports. Senseless and utterly false statements as to the cleanliness and wholesomeness of American meats has had a direct influence, as I shall show in more detail later, upon increasing the restrictions imposed upon importation of American meats into foreign countries. But first let us examine the relation of the export business to home industry.

To get a clear idea of how this immense export business has come into existence and of what it means to the entire agricultural and live stock raising population, it is necessary to go back some distance in the history of our country and note the various changes in the growth of the cattle and live stock industry.

Previous to the development of the packing business in Chicago there was nothing but a local market for beef cattle in any part of this country. The influence of the packing business in broadening the market until now it embraces the whole world has been clearly set forth in a recent work undertaken by authority, and under the auspices, of the National Live Stock Association. This history of the live stock industry treats particularly of the development of cattle raising in the West, taking Texas as the starting point, because Texas was the place where cattle raising on a large scale, as we know it to-day, had its birth.

Texas became known as the great cattle state in the '50s. In their efforts to find more than a local market, Texans undertook the experiment of driving cattle all the way to the new California gold fields, where the food supply was short. They also shipped beef steers by steamship to New Orleans and to Mobile, which were then developing something of a live stock market. In the later '50s they began reaching out for northern markets, driving their herds into Missouri, sometimes as far as St. Louis, and once as far as Quincy, Illinois. Ten and fifteen dollars the head was considered an excellent price in those days for choice selected beef steers.

The panic of 1857 gave a severe setback to the cattle business of that time, and the war, which came on before the country had recovered from the panic, continued to keep down this industry in the south and southwest. It commenced to revive immediately after the war; apparently stimulated, as nearly as the records of those days indicate, by the period of railroad building which set in soon after the war was over.

It is interesting, in view of what has been said in recent days about cattle prices, to compare the prices of now with the prices realized then, when returning prosperity was giving cattle raising one of its first booms. A writer of the period tells how a friend went to a herd of 3,500 beeves and bought 600, taking his pick of the lot at $6.00 the head. Then he took his choice of 600 out of what remained at $3.00 the head. The average price paid for 1,200 picked beeves, selected from 3,500, was $4.50 the head. But Texas was still raising too many cattle for the limited market of that time. The ease with which cattle could be pastured all the year round on the open ranges in a section that had little or no snow encouraged such a multiplication of herds that there was practically no profit in the business. Some of the Texas ranchers, even that early, adopted the method of driving their herds north until they reached a railroad and then shipping to Chicago. But Chicago was only a local market at its best and could not take care of much more than the natural inflow of cattle from the surrounding territory in Illinois and other nearby states.

The cattleman of the southwest, on reaching Chicago in the early '60s, had to sell at prices that were little if any better than he could have got hundreds of miles nearer home; or he could make arrangement with the Chicago slaughter houses to kill and pack his stock on his own account. Packing at that time meant converting cattle into salt or barreled beef. The by-products all went to the slaughter house. This method of disposing of his cattle was usually the least satisfactory of all to the southwestern cattleman.

The next step was the opening of the cattle shipping market at Abilene, Kansas. The old Kansas Pacific Railroad, now a part of the Union Pacific, extended its line late in 1867 to Abilene and that place immediately became an important shipping point. In the fall of 1867, immediately after the Abilene market was opened, 35,000 head of cattle were disposed of there, which was a large number for that time.

But the cattleman was not yet out of the woods. There was a prejudice against beef from the ranges, which was all classified as Texas beef. Early adventurers into the business of shipping cattle East from Abilene found this out to their sorrow. It is related that the second shipment that ever left Abilene consisted of some 1,900 head of beef cattle which had been bought at about $20.00 per head and shipped to Chicago. There was no market at Chicago, so the owner took his cattle on to Albany, which was then an important eastern market. There he sold his herd for $300.00 less than he had paid in freight. The venture stood him a dead loss of the $17,500.00 he had paid for the lot and $300.00 more paid for freight than he received,—a total of $17,-800.00 on the loss side of his ledger, to say nothing of his time and his own expenses.

By 1870 the demand had picked up a little and that year it is related there was a good market at Abilene, good beeves selling for around $20.00 a head. The market was helped somewhat, probably, by a rate war between the railroads, during which freight on live cattle went down to $1.00 a head from western points to Chicago and nothing from Chicago to New York. But there was a reaction in 1871 and higher freight to pay, with cattle still coming in from the southwest in response to the stimulation of the year before. Out of the experiences of that year came the practice of sending cattle from the south and southwest to the northern ranges to be finished and rounded out for the market.

When the Jay Cooke failure and the panic of 1873 came along, the bottom dropped out of the cattle market. The two following years, 1874 and 1875, were also very dull. Conditions were reflected in the records that have come down to us of the greatly reduced number driven north from Texas for further feeding. From more than 400,000 head sent north in 1873 the number dropped to 155,000 in 1874 and to 150,000 in 1875. These were entirely beef cattle practically ready for the market and good stock for further feeding.

From this point we are able to trace definitely the influence of the packers and their methods of doing business upon the live stock industry and the broadening of the market for the product of American farms and ranges until these markets now embrace the entire world. The early '70s were the period of experimentation by the packers with refrigeration in the handling and packing of meats all the year round, and with the refrigerator-car as a means of transporting fresh beef to far distant markets.

The packers had developed the refrigerator-car experiment to the point where they knew they had something that must be used if they were to extend their business beyond a merely local field. They appealed to the railroads, as I have indicated elsewhere, and the railroads refused to undertake the expense, fearing that their cars would lie idle at times if they were required to furnish cars specially adapted to the transportation of fresh meats under refrigeration.

The packers then undertook to build their

own cars, because they could not afford to have this form of transportation furnished in any haphazard or half-hearted way. They foresaw not only that they must have a certain car supply for their own protection, but that the public would be best served and would give them assurance of continued business only by having such a constant and reliable car supply. So the packers built their own refrigerator-cars. That step on their part marked not only the beginning of the private-car lines, but the beginning of the live stock industry as a safe business. *It was the step* as recognized now by historians of the live stock industry *that changed stock raising from an adventure to a business.*

This departure also changed the entire character of the development of the West. The "wild and woolly" characteristics of the plains country gave place to law and order and settled ways of doing things and the Great American Desert began then and there to disappear. Men who are not yet old can remember the days when their school geographies showed the immense expanse of territory extending from southern Texas to Canada between the longitude of the Missouri River and the Rocky Mountains as a desert.

There used to be acrimonious debate as to whether the Sahara of Africa or the Great American Desert was the more barren. Many Americans, determined to yield first place in nothing for their country, proudly insisted that, for barrenness and general worthlessness and for territorial extent, the Great American Desert had the Sahara absolutely crowded out of competition. It was known that buffalo grazed to some extent in the plains country, but the adaptability of that section for any other purpose than buffalo grazing was not realized or even suspected by any except a few Texans who had pastured on its edges.

The enterprise of the packers following

the inauguration of modern refrigeration methods so extended the market for cattle that the cattlemen little by little commenced to invade the Great American Desert. As a direct consequence, practically all of what was then considered a desert now has considerable value and most of it is worth from $30.00 to $60.00 an acre and is still rising in value as good farm lands.

The development of the export business naturally followed the introduction of the refrigerator-car into the packing industry and its effects upon live stock raising and the marketing of meats. The packers had previously met with considerable success in marketing barreled beef, salt meats and smoked meats abroad, but the growing beef business and its possibilities were an additional call upon their ingenuity and enterprise as merchants. They could easily foresee, as clear-headed business men, that extra stimulation of cattle raising and the natural exigencies of the business might easily result at times in a surplus of cattle or beef products that would have a depressing influence upon prices and the business generally if some outlet for the surplus were not found.

The keen competition of the business also prompted them to look for markets wherever they could be had. This competition was probably sharper than it has ever been in any other industry developed in this country. The men who had been attracted to the business were giants of their kind—the stamp of men who have done all of the pioneering and developing in this country at which we now marvel. They could see the immense possibilities opening before them and naturally each of them was ambitious to be first in the race. It followed that they brought to bear all of their tremendous energies and great ability to broadening the field of their operations.

This condition also led them to devise new ways and means of so handling the products of their packing houses that they would find a readier market and of *utilizing to the last degree the merchantable possibilities of the by-product.* These necessities of their situation inevitably drove them into the foreign markets.

The effect of all this pioneering by the packers upon the cattle raising industry of the country is frankly acknowledged in the history to which I have already referred, compiled by authority of the National Live Stock Association. That work, in detailing the influences that operated in the early '70s to bring about new and improved conditions in cattle raising, says:

"But the principal influence that was at work indirectly in behalf of western cattlemen at that time was the development of new features and new methods in the packing house industry. Heretofore the markets for fresh beef from these sources had been, in the main, local in extent and much of their beef output was in the form of salt cured products.

"Exportation of beef on the hoof slowly but steadily was attaining greater magnitude at that time, but it was so hampered by foreign fears—real or pretended—of various infections being introduced into Europe by American cattle, and also by agitations there in favor of home production, that it became necessary for our people to devise other ways and means of getting American beef into European markets.

"As invariably happens in our country, when an imperative demand arises for new ways of doing things, somebody steps forward promptly and points out the necessary effective means by which to do them. In this case the packing house interests quickly solved the problem by sending the foreigners prime dressed beef carcasses that were above criticism or objection; and with these went corned beef, and as the new methods further were developed, a variety of other canned and potted beef products.

"New vehicles of transportation having been required for the dressed beef trade, they came forth without delay in the form of refrigerator-cars on the railroads, and refrigerator apartments in the ships. With these, the packers at Chicago, Kansas City and other great market centres were enabled to deliver beef carcasses on the farther side of the Atlantic in as perfect condition as that in which they were placed upon the blocks of retailers within sight of the packing houses; and with these cars to extend their home trade in dressed beef to every part of the country accessible by railroad.

"This new branch of the packing house industry, which, within a few years later, became by far the largest part of it, made its influence felt strongly, and in 1876 and in 1877 had risen to greater proportions.

Its magnitude in 1878 was reflected in the fact that nearly forty per cent. of all the live stock marketed in Chicago during the year, or about 500,000 head, went to consumers in the form of dressed beef from the packing houses of Chicago. At Kansas City and other packing house centres the dressed beef industry held about the same ratio to the total number of cattle put upon their markets."

But the opening of foreign markets and development of the packing industry of Chicago did not all at once put the cattle raising and meat producing industry on the basis that we know now. The grade of cattle and the grade of beef produced in the '70s was still not of the very first class and beef was not so universally eaten as it is now. This condition was reflected in the prices that prevailed at that time. In 1876 and 1877, $2.75 to $3.00 per cwt. was considered a good price for beef cattle.

Some Texas men thought they had a

"nice market" when fat beeves from their ranch corrals could be sold around \$2.25 per cwt. Other records of the period give \$2.50 and \$3.00 per cwt. as considered an excellent price. A record of top prices by months for northwestern "rangers" in the summer of 1878, except during August, shows that the very highest prices ranged from \$3.75 to \$4.40 per cwt. "Texas" cattle during the same period sold for \$3.00 to \$3.90 per cwt. The average price for "rangers" during a six months' period was only \$3.95 and for "Texans" \$3.55 per cwt.

At Albany, in that year, extra cattle weighing from 1,250 to 1,300 lbs. brought only \$5.00 to \$5.37½ per cwt. and this had to cover freight from the far western ranges. Dressed beef of the same extra quality sold at \$10.50 to \$10.75 per cwt.

By comparing these price quotations with prices to-day, we have a striking illustration of how development of the packing industry, with its utilization of by-products, and its foreign market finding has reduced at both ends the margin between live cattle prices and dressed beef prices, making live cattle prices higher and making dressed beef prices lower.

In 1878 we find the wholesale price of extra beef, dressed, brought down from \$13.00 and \$13.25 in 1874 to \$10.50 to \$10.75 a cwt. The cattle that produced this \$10.50 to \$10.75 beef sold at \$5.00 to \$5.37½ a cwt., so there was still a difference of about \$5.50 a cwt. between beef cattle on the hoof and the same cattle dressed. To-day, as the daily market quotations show, the price difference between live beef cattle and the same cattle dressed is only about \$2.00 a cwt., and frequently it is less. This elimination of the wide margin between live and dressed beef prices is due entirely to the saving effected by the packer in utilizing the by-products, and finding a foreign market for meats that otherwise would be handled at a loss, and so would make it necessary to ask higher prices for the meats consumed in this country.

It is easy to study this phase of the subject in the market quotaticns published from day to day in the newspapers. The packer sells at wholesale the dressed beef from every steer slaughtered for much less than he paid for that steer on the hoof. He must pay all expenses, get all of his profit, and make up the difference between live cost and dressed selling price out of the by-product materials, most of which were wasted before the early packers applied their genius to the business.

It is obvious that the slaughtering and distributing of beef if undertaken by any agency less perfectly equipped than the packers are, would necessitate very much higher prices to the consumer than are now paid. Commissioner of Corporations Garfield's report shows that the packers get less than a dollar a head out of the cattle handled. To do this they must utilize every scrap of by-product to the last degree, and then find a market for it in every part of the world.

The influence of the packers, and their development of the export trade, along with the domestic business, became pronounced about 1884 and 1885. The year 1884 was the first in which one million cattle were slaughtered in Chicago. Since 1890, at times, nearly two and one-half million head of cattle have been slaughtered in Chicago in a year, and in many years since 1890 from one million to a million and a half of cattle have been received at Chicago and reshipped on the hoof, more cattle being shipped out on the hoof than the entire receipts amounted to before the packers became a factor in the market.

The great changes in the industry have come since 1885. The tremendous development, not only in the packing industry itself, but in the raising of stock and the new and improved methods in both branches of the industry, have nearly all come within that time. Speaking of this period of revolution, the same official history of the live stock business which I have previously quoted has this to say:

"*What had been an adventure was converted into a business.* The rearing of live stock on the western plains and in the mountain valleys ceased to be a reckless 'chancing' of things, ceased to be a speculation, and became a careful, systematic, commercial enterprise, seeking regular and reasonable returns for its output of beef and mutton, which were made to be of the highest grade. The typical owner of great herds of cattle no longer was distinguished by a broad-brimmed hat, no longer stuffed his trousers into the tops of high boots, no longer wore accoutrements suggestive of wildness and wooliness, but presented the appearance of a conservative, unostentatious man of business. He was not out of place, but was at home and at his ease wherever he might be, whether in the great market cities or at his ranch—very much a man of the business world."

The factors in working this change, according to the authority I have mentioned, were improved stock, prudent management and individual control of more or less land upon which each stockman operated, accompanied by the use of fences. These factors were given a chance to operate by the work of the packers. It was the packers who created a great central market, who devised ways and means of disposing of the dressed beef and the beef product, and who opened up foreign markets to take care of surplus ranch products.

It was their work, I repeat, which gave the live stock industry *an opportunity* to be converted from an adventure to a business. They encouraged the growing of quality. They showed the western and southwestern ranchers that the long-horned Texas steer was neither the best nor the most profitable animal to be grown for market. They gave the impetus to improving the grade of range cattle. The ranchmen commenced spending a part of their profits for blooded bulls and crossing them with the native cattle. They commenced to raise stock for meat and not for tallow. This phase of the change has been commented upon in the official history I have quoted as follows:

"The best, and therefore the high priced beef, lies along the animal's back, and anyone can understand that a broad-backed steer that has utilized its food to increase its aggregate of sirloin and porterhouse parts is far more valuable than the narrow-backed, slab-sided animal, perhaps of nearly the same gross weight, but which has utilized most of its food in the production of tallow. The western cattleman saw this and began to produce, with the same amount of food, beeves that yielded the high priced steaks worth from 15 cents to

25 cents a pound in a normal retail market, instead of tallow and medium or low grade meats, worth whatever a buyer could be persuaded to pay for them."

From these changes in the methods of raising cattle and marketing them and their products have come the changes in the character of the range country. The western cattleman's home place, while it is still called a ranch, now partakes more of the nature of what we are accustomed to think of when we say farm. He does not depend upon letting his cattle run out of doors all the year round, finding their food where they can. He provides shelter for stormy weather. He plants and harvests crops that will serve as fodder in the short grass season; he cuts hay, and stores it against the time of grass shortage.

The mowing machine is now as prominent a part of the furniture of the average cattle ranch as the "chuck-wagon" used to be. Small holdings are taking the place of the large ranges. While there may not be so many "cattle kings" as there used to be in the business, there are innumerably more men who are making money at it.

Along with the revolution in the production of beef cattle which I have sketched, there has come great improvement in facilities for getting the cattle to market, for feeding and watering in transit, and so on. These changes have had a marked influence upon the allied business of feeding or finishing cattle for market.

Stock cattle are taken from the big grazing ranches and are turned over to smaller farmers further north and east, who feed and finish them for the market, and so enable us to get the prime beef which we are accustomed to have upon our tables. This is a change that has brought real prosperity to a great part of the western farming community, particularly in corn raising states like Illinois, Iowa, Kansas, Nebraska and Missouri. Instead of selling his corn for 25 cents or 30 cents a bushel, the farmer now puts this corn into beef cattle and gets 40 and 50 cents a bushel for it; and the farmer who does not want to feed cattle, but has corn, also gets a higher price because there is not the same relative quantity of corn going into the market.

These same changes in cattle growing have helped to give a little additional profit to industries as far removed from meat producing as the cotton industry. Cotton-seed that used to be waste is now utilized, in part at least, in cattle feeding enterprises.

That the pioneering and foreign market hunting efforts of the early American packers were great factors in bringing about these changes in American industrial life—changes that were and have continued to be beneficial to hundreds of thousands of American people—is evidenced by what we read when we look back concerning the introduction of American meats into foreign markets.

An English paper of twenty years ago had a paragraph to the effect that "prime Scotch cattle and Highland mutton are as drops in the bucket compared with the enormous imports from America." It said in the same article that a Mr. Henry S. Fitter had disposed of, to retail butchers, $12,000 worth of American meats in one day; a consignment from the Chicago packers.

This meant that the meat so sold came from cattle which had been grown on western farms or ranges, brought to Chicago, and there converted into beef and then marketed four thousand five hundred miles toward the other side of the world. The American farmer received the benefit of this enterprise.

But the packer of that day, who was doing his utmost to extend the demand for American farm products, did not have easy sailing and complimentary words any more than he has to-day. Both at home and abroad he had to fight his way every step he took. At the great London beef markets, as elsewhere, the salesman tried to shut out American meats, but, of course, American enterprise was equal to obstacles of that kind.

The Americans had meat to sell and were there to sell it. If they could not sell it in one place or one market they would make a market of their own. They did so and they did it without resorting to any subterfuges, or attempting to palm off American meats as "home grown" to avoid discrimination. They were out to make a market for American meats, and they sold them for just what they were, knowing that the quality would create a demand.

Here at home, when the successful innovations of the packers commenced to attract attention, they had to run the gauntlet of hostile legislation that was even more troublesome, in some respects, than what they are threatened with now. The rapid development of cattle raising had brought on the cattle boom of the early '80s, which I have mentioned elsewhere, and, of course, that boom had its inevitable disastrous reaction. The packers were given the blame, of course, for the low prices which prevailed in the late '80s and which were due to nothing on earth except an increase in the supply of cattle that far outstripped the demand and all the energy and resourcefulness of the packers in building new markets.

The year of 1889, as we look back, was remarkable for the efforts made in widely separated parts of the country to hamper the development of the dressed beef and packing business by hostile legislation.

Virginia made a law that no fresh meat should be offered for sale at any place one hundred miles from where it had been slaughtered, unless inspected alive by the local inspectors. The inspection fee ranged from \$6.00 to \$10.00; a prohibitive price, of course.

Topeka tried by ordinance to shut out the dressed beef of Kansas City.

Minnesota enacted a law providing that no beef should be sold within any municipality of the state unless inspected by the local officers within twenty-four hours before slaughter.

Colorado undertook to prohibit the sale of beef, mutton or pork unless inspected on the hoof before slaughter.

Indiana's law required that beef, mutton, veal, lamb or pork had to be inspected alive in the county where it was offered for sale.

Of course none of that legislation stood the test of review in the courts. I mention these cases merely to illustrate the obstacles which the packing industry has had to fight, and fight alone, while it has been opening and expanding markets for the products of the ranch and farm.

What the Chicago packers accomplished

in the face of such opposition is best illustrated by a little comparative figuring.

In the twenty years from 1880 to 1900, the increase in population of the United States was in round numbers fifty per cent. During that same period the number of beef cattle marketed at the four principal markets increased five hundred per cent. I know of no other lines of business dealing with natural products—cotton, wool, grain—in which such a marvelous increase in the quantity of material to be handled has been so successfully coped with.

But the packers seemed to be equal to any emergency, and in spite of one or two comparatively low-priced periods, which were due to perfectly natural causes, they kept the market in a healthy condition; and some stock raisers, the careful business men among them, made money practically all the time.

When we talk about low-priced cattle and of farmers losing money in their cattle ventures, is it not fair, instead of blaming it all on a "beef trust," to remember that raising cattle is a business like selling shoes or dry goods or making chairs?

Not nearly all of the people who venture into these various other undertakings make money; statistics show, I believe, that the great majority that go into business meet with failure. I think you will find that the percentage of those who lose money in raising or feeding cattle for market is smaller than in almost any other line of producing effort. I believe, too, that wherever you find a man in the cattle business who will class as a good business man you will find one who makes money almost every year, and one who, taking into account a reasonable period of years, makes a profit that is entirely satisfactory to him, figured on the basis of his investment.

In this connection I am moved to quote a few words from an article by Mr. Samuel W. Allerton, written a short time ago. Mr.

Allerton was one of the early pork packers and cattle dealers in the Chicago market. He was in the business quite a long time, but now, for several years, he has been extensively a farmer, owning, I believe, the largest farm in Illinois; and a large part of his farming business revolves around the feeding and finishing of cattle for market. In an article for a live stock paper Mr. Allerton said:

"I remember when I first went into the business, in 1853, New York consumed about three thousand two hundred cattle per week, Boston about five hundred, Philadelphia five hundred and Baltimore about two hundred. There were not many more than five thousand cattle transported per week. Now we transport at least one hundred and twenty-five thousand over our railroads, so you see the cattle industry has increased very largely over our population, and we must give the dressed beef men credit for finding a market for this great surplus, as they have hunted a market for it in every spot on the globe."

As to the farmer's point of view on this subject, Mr. Allerton said: "Being a farmer, I know how they feel. They, of course, always want a little more, and every little newspaper in the country thinks it is smart to howl about the dressed beef 'trust,' when the truth is there is no body of men which has done so much to promote the live stock interests of this country as the dressed beef men. They have put fresh beef into every state in the Union where they were not accustomed to eating fresh beef, and they have hunted the world over to find a market for the coarse cuts of beef. Every market which they are in has steadily grown and increased, which is evidence that they have been beneficial to the live stock interests of this country."

CHAPTER XIV

THE GROWTH OF PORK PACKING

WHAT I have said as to dressed beef and live stock raising, and the influence upon the industry exerted by the packers in their discovery of new methods of marketing and of new markets, applies with equal force, in practically every detail, to all the other branches of raising meat animals and converting them into food for the table. The growth of pork packing in the West illustrates what I mean.

Pork packing was a considerable industry before the dressed beef business came into existence on any large scale. The very earliest packers purchased live hogs at many different points in the West, slaughtered and cured them and shipped the product to all parts of this country and to foreign countries. In the period between 1880 and 1885, before the dressed beef business had attained proportions at all remarkable, pork packing was already an important business. During that period the western packers slaughtered fifty-two million, three hundred and twenty-two thousand hogs, an average of ten million, four hundred and sixty-four thousand, four hundred hogs a year.

Now the average swine raiser probably has not noticed any particular increase in the raising and marketing of hogs during recent years. Yet, during the last five packing seasons ending with March, 1906, pork packing at western points absorbed very nearly one hundred and eighteen million hogs, an average of twenty-three million six hundred thousand hogs a year. This immense increase in the production of hogs has been handled—purchased, slaughtered and marketed—without a break or a material change in the market; in fact, my recollection is that the price has been going up practically all of the time.

The handling of this immense increase in hog production is due entirely to the business improvements brought about by the large packers in their dressed beef and kindred businesses. This would seem to be so from the fact that there has been no material increase in hog marketing or pork packing at any of the markets where the beef packers are not doing business.

The methods of the packers, due to refrigeration and transportation facilities resulting from their private-car enterprises, have made hog marketing and pork packing an all-year-round industry instead of a winter industry as it used to be. This, of course, is an advantage to the hog raisers, not only because it gives them a market all the year round, but because it has brought about the absorption of practically double the number of hogs.

To show this more clearly, let us glance

at the figures showing the number of hogs packed at Chicago and classified as winter packing and summer packing. The following table shows that while winter packing at Chicago has remained almost stationary in the last twenty years, summer packing has caused the total pack to be almost double:

HOGS PACKED AT CHICAGO

Year.	Summer Packing.	Winter Packing.
1882-83	1,664,957	2,557,823
1883-84	1,900,408	2,011,384
1884-85	1,859,988	2,368,217
1885-86	2,535,678	2,393,052
1886-87	2,581,752	1,844,189
1887-88	2,000,741	1,731,503
1888-89	1,774,228	1,429,723
1889-90	2,294,027	2,179,440
1890-91	3,211,144	2,908,418
1891-92	2,498,754	2,706,284
1892-93	2,873,883	1,478,212

1893-94	2,523,587	1,695,980
1894-95	2,817,734	2,475,468
1895-96	3,114,940	2,375,470
1896-97	3,684,220	2,283,375
1897-98	4,074,535	2,672,730
1898-99	4,767,290	3,249,385
1899-1900	4,249,860	2,869,580
1900-01	4,298,420	2,970,095
1901-02	4,202,095	3,433,905
1902-03	3,908,260	2,952,193
1903-04	3,787,126	2,925,960
1904-05	3,177,842	2,812,588
1905-06	3,545,197	2,592,866

Chicago, of course, has no monopoly in pork packing. This is an industry that is widely distributed throughout the West. The following table shows the number of hogs packed in the West for the twelve months ending March 1st, 1906, at the fifteen leading points (the others being lumped) with comparisons for previous years:

Point.	1905-06.	1904-05.	1903-04.	1902-03.	1901-02.	1900-01.	1899-1900
Chicago	6,138,063	5,990,430	6,713,086	6,860,453	7,636,000	7,268,515	7,119,440
Kansas City	3,108,172	2,890,962	2,086,550	2,055,942	3,427,802	1,981,288	2,621,727
South Omaha ...	2,197,295	2,054,464	2,173,734	2,004,826	2,390,416	2,241,599	2,192,496
St. Louis	1,777,657	1,908,592	1,570,744	1,262,358	1,725,407	1,566,550	1,507,951
Indianapolis	1,447,250	1,275,887	1,123,665	930,000	1,225,310	1,185,600	1,145,252
Cincinnati	646,325	643,524	585,873	498,376	569,782	617,032	656,244
Milw' & Cud'y ..	928,706	865,308	846,287	553,986	790,063	911,256	864,590
St. Joseph	1,816,502	1,588,623	1,609,185	1,528,860	2,105,293	1,723,377	1,346,733
Cedar Rapids ...	568,888	495,121	501,789	391,524	496,790	496,308	427,637
Ottumwa	631,217	669,342	486,991	441,680	610,002	653,785	688,939
Cleveland	648,333	678,269	625,803	521,672	496,231	500,785	489,282
Louisville	411,720	404,879	319,969	323,940	375,000	360,425	397,975
Sioux City	999,966	597,516	467,595	777,320	879,763	733,754	514,238
St. Paul	875,625	884,187	810,988	715,327	658,591	514,385	394,093
Nebraska City ..	315,673	268,920	238,707	209,835	180,746	114,962	235,923
Fifteen places	22,511,392	21,216,024	20,160,966	19,076,009	23,567,096	21,869,621	20,602,517
All other "	3,063,368	2,702,399	2,214,720	1,529,562	1,844,580	1,731,053	1,598,304
Total in West	25,574,760	23,918,423	22,375,686	20,605,571	25,411,676	23,600,674	22,200,821

For these twenty-five million, five hundred and seventy-four thousand, seven hundred and sixty hogs slaughtered at western points in 1905-06, the western packers paid the western farmers the enormous sum of $302,487,000, or nearly $12 a head. Eastern slaughterers, during the same period, paid eastern farmers about $65,000,000 for five million, seven hundred and ten thousand hogs, or nearly $11.50 a head.

The tendency of prices has been upward, with, of course, fluctuations due to natural causes, as in 1902, when the short corn crop sent up the price of beef cattle, and sent up the price of hogs to an abnormal figure for the same reason, as well as through sympathy with the higher price of cattle and beef. I am presenting here a table showing the monthly range of cash prices for live hogs, the highest and lowest price of the month, covering a period of eleven years in the Chicago market. The table is as follows:

Month.	1905.	1904.	1903.	1902.	1901.	1900.	1899.	1898.	1897.	1896.	1895.
January	$4.15	$4.20	$5.70	$5.55	$4.80	$4.15	$3.30	$3.35	$3.00	$3.35	$3.70
	5.00	5.20	7.10	6.85	5.47½	4.92½	4.05	4.00	3.60	4.45	4.80
February ...	4.45	4.30	6.05	5.45	5.10	4.55	3.45	3.60	3.10	3.60	3.60
	5.15	5.80	7.55	6.55	5.65	5.10	4.05	4.27½	3.75	4.35	4.65
March	4.55	4.70	6.50	5.65	5.20	4.60	3.50	3.65	3.35	3.55	3.85
	5.55	5.82½	7.87½	7.00	6.20	5.55	4.00	4.17½	4.25	4.25	5.30
April	4.70	4.40	6.60	6.35	5.55	5.10	3.50	3.60	3.50	3.05	4.40
	5.72½	5.50	7.65	7.50	6.25	5.85	4.15	4.15	4.25	4.15	5.42½
May	4.60	4.20	5.40	6.40	5.50	4.90	3.45	3.70	3.25	2.80	4.10
	5.65	4.95	7.17½	7.50	6.00	5.57½	4.05	4.80	4.05	3.75	4.97½
June	4.50	4.35	5.50	6.65	5.55	4.85	3.45	3.55	3.10	2.70	4.20
	5.65	5.45	6.50	7.95	6.30	5.42½	4.00	4.50	3.65	3.60	5.10
July	5.00	4.75	4.75	7.00	5.30	4.85	3.55	3.60	3.05	2.50	4.50
	6.17½	5.90	6.10	8.25	6.40	5.55	4.70	4.17½	4.00	3.70	5.70
August	5.25	4.55	4.50	6.10	5.40	4.75	3.85	3.45	3.45	2.50	3.85
	6.45	5.80	6.20	7.00	6.70	5.57½	5.00	4.20	4.55	3.70	5.40
September ..	4.85	4.70	4.85	6.80	5.80	4.85	3.90	3.40	3.60	2.45	3.55
	6.20	6.37½	6.40	8.20	7.40	5.70	4.90	4.15	4.65	3.50	4.65
October	4.35	4.55	4.55	6.00	5.45	4.30	3.80	3.25	3.20	2.55	3.20
	5.80	6.27½	6.50	7.92½	7.10	5.55	4.90	4.00	4.40	3.65	4.50
November ..	4.35	4.15	3.85	5.75	5.00	4.35	3.55	3.10	3.15	2.90	3.20
	5.25	5.25	5.50	6.95	6.30	5.12½	4.35	3.85	3.80	3.70	3.85
December ...	4.55	4.10	4.05	5.60	5.10	4.40	3.55	3.15	3.10	2.90	3.25
	5.35	4.87½	4.90	6.90	6.90	5.10	4.45	3.70	3.60	3.60	3.70

This tremendous growth in hog raising, in consequence of the improved business methods and foreign market-finding by the packers, has been as widespread in point of benefits conferred as the growth of the stock raising industry. It has had its effect upon every farmer who raises a few hogs, not only throughout the great West, but in the more thickly populated eastern states where hogs are raised exclusively for the local market.

The prices obtained in Chicago and other packing centres have naturally made better prices in the small towns. The hog crop, too, has been another outlet for the corn crop of the farmer of the middle West. As in the case of feeding and finishing cattle for market, the cattle raiser can feed corn to his hogs and get more for it than by marketing it as corn; and a considerable and valuable part of his corn goes back upon his own land, as I have elsewhere observed.

This is an important point in connection

both with cattle raising and hog raising, by no means to be overlooked in estimating broadly the benefits brought to the agricultural sections of the country by the meat industry. The farm of one crop, or many crops, when devoted exclusively to growing cereals or cotton, steadily deteriorates. The live stock farm, whether its specialty be cattle, hogs or sheep, improves from year to year.

Hog raising, too, in many parts of the country, particularly in the Mississippi Valley and further west, has been developed as an absolutely new crop, a new source of wealth to the farmer. This has been the case in Illinois, Wisconsin, Minnesota, Iowa, Nebraska, Kansas, Missouri, in fact all through the Mississippi Valley. In all those states hog raising has been stimulated and made more profitable, and in fact has been given an excuse for existence, by the establishment of packing plants at Kansas City, St. Joseph, Omaha, Sioux City, South St. Paul, and Fort Worth.

This phase of the subject can be best studied right now in connection with the packing business at Fort Worth, because there it is in its beginnings. Two large packing establishments have been in operation in Fort Worth about three years. Already their influence upon the live stock interests and especially hog raising in that part of the country is very noticeable.

The Texas hog of a few years ago was a cousin, as it were, of the old-time Texas long-horned steer. He was what is commonly known as a razor back, long of leg, fleet of foot, sharp of nose, and nothing to brag about as a producer of meat. Within three years there has been a noticeable change in the character and quality of the Texas hog. The farmers there have begun to improve their breeds, and to grow for more weight and better quality. They are doing it because the packers have given them a market at their very doors.

The Fort Worth market also shows, in plain figures, how the packing industry, with its worldwide effort to find new markets, directly promotes the live stock industry. *Every extension of the packing industry makes the raising of meat animals a profitable business for an increased number of American farmers.* This has happened in the territory tributary to Fort Worth. We see it the more clearly there because that market is so new and is established in territory that has been the cradle of stock raising on a large scale.

Last year, with other markets holding their own, the Fort Worth market showed a larger increase in the actual number of cattle handled than any other market in the country. Its percentage of increase in cattle receipts was more than three times that of any other market. And the number of hogs received there was even higher proportionately. Note these figures:

Receipts of cattle and calves at Chicago in 1905 were 176,000 more than in 1904, an increase of about five per cent.; Kansas City receipts increased 186,000, a little over nine per cent.; St. Louis increased 48,000, a little over four per cent.; Omaha increased 86,000, or a little over nine per cent.; St. Joseph decreased slightly.

Fort Worth receipts of cattle and calves increased 226,182 over 1904 receipts—a *gain of more than thirty per cent.*

The hog showing is even more remarkable. At Chicago hog receipts for 1905 increased 533,189 over 1904, a gain of nearly seven per cent.; Kansas City receipts increased 976,615, or a shade under 44 per cent.; Omaha fell off a trifle; St. Louis increased 71,413, about four per cent.; St. Joseph increased 245,037, or almost 15 per cent.

Fort Worth receipts of hogs in 1905 were

186,592 more than in 1904, *a gain of more than 60 per cent.*

The striking feature of these figures is that other markets were practically holding their own while Fort Worth was making its phenomenal gain. All of them, except St. Joe in cattle and Omaha in hogs, made a gain in spite of Fort Worth. The only rational explanation of this is that a great part of the cattle and hogs marketed at Fort Worth represented *increased production* by old live stock raisers and *absolutely new ventures* into the business by farmers who did nothing with cattle and hogs until the Fort Worth market was created. In this respect Fort Worth has but repeated the history of other markets—Chicago, Kansas City, Omaha and the rest.

What has been said of cattle and hog raising generally applies with equal force to sheep raising. The foreign market for mutton and mutton products is not large as yet, but the foreign market provided for beef and pork gives mutton a chance in the home market. Receipts of sheep at the Chicago Union Stock Yards have been quadrupled in twenty years—have risen from 1,008,790 in 1886 to 4,736,558 in 1905. Receipts of sheep at Omaha have been multiplied by ten within fifteen years—increasing from 170,000 in 1891 to more than 1,700,000 in 1905; at Kansas City and St. Louis, during the same period, receipts of sheep have been more than trebled. All this means that the income producing power of the West has been increased just that much

This immense increase in cattle raising and hog raising, as I have outlined it, due to the opening up and settlement of the great West during the last quarter of a century, would have been, in some aspects, a calamity if it had not been accompanied by corresponding increase in outlets for the product. Consumption of meats has been greatly increased throughout the country by the distributing facilities and the economies which the packers have furnished, but production has been far beyond the consuming power of the United States. Yet, as we have seen, the tremendously increased production has been, on the whole, absorbed without a material break in prices of the live meat animals from the farm and range.

What has made this increased absorption possible? *The export business;—nothing but the export business;—the export business which the packers have developed* in the face of violent and unreasonable attacks both at home and abroad. We see this clearly when we look at the export business in comparison with the gross domestic business in live stock.

The live cattle exported, together with the exports of fresh beef and beef products, as shown in the tables I have given, represent in the aggregate at least one million head of cattle—or the value of one million head annually sent to foreign markets. This number is equal to one-fifth of the total annual slaughter of cattle at the four greatest western markets—Chicago, Kansas City, South Omaha and St. Joseph.

The exports of hogs, pork, bacon, hams, lard, and other hog products last year represented, in the aggregate, the market value of nearly ten million live hogs delivered at American markets. This number of hogs would be equal to two-thirds of the hogs slaughtered at the four greatest western markets—Chicago, Kansas City, South Omaha and St. Joseph.

Think of the effect upon prices to the producer and upon market conditions generally, if this immense outlet for American farm products did not exist! The foreign markets, opened and developed by the unaided efforts of the packers, have taken care of this enormous surplus, have kept the home market steady, and have enabled the American farmer, cattleman and hog raiser to make a profit on his live stock, and on the grain and farm products he feeds them.

The export business in cattle and meats and meat products is the safety-valve of the stock raising and meat producing industry. It takes care of the surplus from the farms. It provides a market for grades of cattle and beef that would find but a limited market, or none at all, in this country. It thus preserves the balance between production and consumption, giving the American stock grower a fair price for his product, and giving the American consumer the kind of meat he wants at a fair price.

But large as this export business is, it could be made much larger, with corresponding increase in profit to live stock raisers and meat producers, by a little consistent and co-operative effort in this country.

Great Britain is now our best customer. Her markets absorb about $150,000,000 a year of American feed lot products. But

Great Britain is an outlet for the better grades of live stock, leaving us to look elsewhere for a market for the commoner qualities.

Continental Europe ought to—and would, under proper cultivation—furnish a vast outlet for the classes of meats for which there is least demand in America and Great Britain. The masses of the people in Continental Europe are practically without beef at all times. Horse meat is a recognized article of commerce over there. They would be delighted with cuts that our people pass by. Instead of the steaks and roasts which we demand, they would be pleased with boiling cuts and corned beef. As for other meat products—hams, bacon. sausage, etc.—a market for literally millions of American corn-fed hogs would be provided in Germany, France and other Continental European countries if it were not for the restrictions upon imports from America, *which restrictions are encour-*

aged by the unwarranted attacks by Americans upon an American industry.

The possibilities of this foreign market are emphasized by a statement prepared a little over a year ago—and conditions have not materially changed since—by one of the most careful statisticians in the country. Here it is:

"Great Britain's flocks and herds have, if anything, gone back, while the population has gone ahead. The flocks and herds of Ireland have virtually been stationary for a quarter of a century. Twenty-five years ago France had one hundred and eighty-eight head of live stock per one thousand acres of her area. She now has one hundred and sixty-four head, or a dead loss of twenty-four head per thousand acres.

"Germany, a quarter of a century ago, had three times as many sheep per head of population as now. The Fatherland now has fewer cattle per capita than then.

"Holland and Switzerland have only

one-half as many sheep per head of population now as they had two and a half decades ago, and Belgium only a fourth as many.

"In these countries, during that time, the flocks of sheep alone have actually decreased from one hundred and four million to seventy-five million, showing an actual loss of twenty-eight per cent. At the same time the increase in the population was twenty-five per cent., thus making the comparative loss much greater.

"The combined population of Germany, France, Austria-Hungary, Switzerland, Sweden, Denmark, Belgium and Holland twenty-five years ago was one hundred and forty million. This human family has now increased to one hundred and seventy-five million people or twenty-four per cent. Meantime, the combined herds of cattle of these countries have only increased from forty-eight million to fifty-eight million head or about twenty per cent.

"The relative scarcity of meat upon the continent is readily seen by the excessively high prices paid there for all carcass meats. These prices average one hundred per cent. higher wholesale than similar meats sell for in the United States, and thirty to fifty per cent. higher than they do in England, even in the face of the facts that labor and other continental items of production are cheaper than in either Great Britain or the United States.

"The growing scarcity of the world's edible meats is produced by the two causes previously given: the faster increase of the human race in proportion to the increase of abattoir animals, and the improved condition of the working classes, which causes a greater per capita consumption of meats than existed two decades ago.

"The per capita consumption of meats in the United States has increased fully twenty-five per cent. during the last fifteen years. In Great Britain it has nearly doubled in the same time. On the continent, the demand, as measured by the very high local price of meats, has largely increased, but the increased consumption has been limited by the virtual exclusion of foreign meats and the insufficiency of domestic herds to supply the local demands. With improved industrial conditions and continuing high tariffs or other means for excluding the surplus of other countries, the masses of the continental nations must desist from meat eating or pay exorbitant prices for this essential staff of life."

The restrictions which hold back Americans from supplying this tremendous potential market in Continental Europe are of various kinds—tariffs, restrictive regulations as to the form in which meats shall be shipped, unreasonable regulations as to inspection, and absolute prohibition. The tariffs, in addition to their revenue purpose, are frankly intended to restrict meat importations so as to give the home grower a higher priced market. Regulations as to form in which meats may be imported, and as to inspection, in many instances, would seem to be devised, when we come to analyze them, for a similar purpose of exclusion, although their ostensible purpose is to protect public health. The foreign politicians find in the unwarranted attacks by Americans upon the packing industry an excuse for maintaining these restrictions.

We find in Germany a concrete example of the way hostile foreign legislation, encouraged by the anti-packer "bushwhacking" in America, operates against increase in our exports of meat animals, meats and packing house products. In 1904 (the last year for which I have official figures) Germany imported three hundred and twenty-one thousand, eight hundred and seventy-nine cattle worth $26,796,700, mostly from Austria-Hungary, Denmark and Switzerland, less than one-half of one per cent. coming from other countries. During the same year we sold Great Britain four hundred and one thousand, two hundred and forty-five cattle worth $34,844,378.

Given a fair opportunity, there is no reason why we should not be able to sell Germany a large proportion of the more than three hundred thousand head she imports annually on the hoof, to say nothing of the immense quantity of pickled and canned meats her people would consume—would be glad to get—if given a chance. Persons conversant with German conditions and requirements at first hand estimate that we ought to be able to sell Germany two thousand five hundred to three thousand five hundred live cattle every week if given entrance to her markets. When Germany practically stopped taking American canned meats at the close of 1900, her annual imports of those goods represented one hundred thousand cattle a year, about two thousand head a week.

An important point to be considered in

this connection is that the German market would absorb the grade denominated "range cattle," for which there is but a limited demand at home and in England.

Now for a closer examination of the restrictive regulations upon imports of American meats into European countries (taking Germany as an example) and the influence of our American critics upon those regulations. Practically all of the German states prohibit importation of live cattle from America. Those ostensibly permitting it surround importations with quarantine and inspection regulations (each one with a fee attached) that are as effective in barring cattle as flat prohibition would be.

The ostensible reason for these regulations is to protect German cattle against infection from diseased American cattle. As a matter of fact, American cattle are far freer from disease than are German cattle or the cattle of any country from which Germany draws its supply. American cattle are better cared for than are European cattle; they have more room; they run out of doors on ranges and pastures of considerable size, instead of being confined to restricted quarters as so many European cattle are.

Most of what little cattle disease we have comes to us from Europe. Take the much talked of tuberculosis for example. Official reports of United States inspection at the big packing centres show that only about one per cent. of American beef cattle are infected, and these are condemned. Figures from across the ocean indicate that about thirty per cent. have tuberculosis. So it cannot be that fear of disease is all the reason for keeping American cattle out of Germany. But that is a good excuse; and every time one of our "yellow" American writers runs amuck in a spasm of "seeing things" the German politician gets another bit of assistance toward keeping up the bars against American cattle.

American fresh meats are nominally admitted to German ports; actually they are as effectually barred out as if prohibited. German laws provide that fresh beef or pork cannot be imported in "quarters" or "cuts" but must come in entire carcasses with all or part of the head and practically all of the internal organs naturally attached to the carcass. It is impracticable to ship meats in this form.

Since 1900 American canned meats and sausages have been barred absolutely from Germany. There is no exception. "Yellow" imaginings of the dark secrets of packing houses help to retain the prohibition. Yet Germany takes our barreled beef—a quantity of it that represents more than fifty thousand cattle a year. There is no reason why she should take this and reject other meats, except that she must have it, and has not as yet been able to get elsewhere a sufficient quantity of the right quality. And it is a fact that the *German government has bought largely of American canned meats for the Imperial navy and the commercial marine service at a time when importation of these goods for general sale was prohibited.*

What I have said of Germany applies more or less to France and other countries of Continental Europe where we ought to have a better market for the products of American farms. The discrimination against American meats, encouraged by the unjustifiable utterances of irresponsible Americans, has already been felt. For not feeling it more seriously as yet we have to thank a series of unusual conditions throughout the world during the last half dozen years.

The Boer war, the Boxer troubles in China, the Russo-Japanese war, the military and naval precautions arising from "wars and rumors of wars" elsewhere, have served to keep up a market for the "range" cattle that we should be selling in Continental Europe. But we cannot depend upon these extraordinary conditions all the time. When the market drop comes, it will fall heaviest on the American farmer and live stock raiser.

American encouragement of this European discrimination is the more dangerous to American industry because of the rising importance of the Argentine Republic as a cattle and meat producing country. On this point let me quote one of the most prominent live stock men in the West, W. A. Harris, former United States Senator from Kansas. In a public address last December (1905) on ways and means of extending our markets in Europe, he had this to say of Argentine's competition:

"Now one idea that has been in the minds of a great many of our public men for a great many years is that they—Continental Europeans—have got to buy their bread and meat of us, and you cannot starve a nation—they will buy anyhow. This is no longer true. The world has moved on. Today the Argentine Republic is the strongest and most dangerous competitor that we have found for the trade with Great Britain, and they are crowding us closely. It is the greatest natural cattle country that the world ever knew. It is the state of Illinois multiplied by one hundred, when you come to consider it. In the northern part of the country you can grow figs and oranges easily in the open air, and it extends in an unbroken plain to the south for two thousand five hundred miles, and it is all rich prairie land practically.

"They are spending more money, and have been in the last ten years, than any other people on earth in order to improve their stock of cattle. They are lavishing fortunes upon the best bulls and cattle that Great Britain can produce, and they are able to raise cattle that would astonish the American citizen. They also have enormous sheep interests.

"They have learned how to ship dressed beef. For a while we rested secure in the idea that they had to pass through the tropics to reach the continental market, and that they could not ship dressed beef in the chilled condition that was required; and that was a great difficulty for a while. But they came to Chicago and took down the best experts they could get in that line of business, and they have built up packing houses there that now send dressed beef and mutton to Liverpool, and they are cutting our throats in the English market, to say nothing of the continental markets."

Now I do not desire to be misunderstood as contending that all restrictions and discriminations laid upon American meats in Germany and other European countries are due entirely to American attacks upon the packing industry. The primary influence at work in Germany against American imports is the agrarian influence seeking to make higher prices for German farm products. But I do contend that every time a "yellow" writer for "yellow" American magazines and publishing houses makes assertions such as have recently been made—assertions which cannot be justified either as fact or as clean fiction—he does his utmost to bring about conditions that will be disastrous to the American stock raiser and farmer, and to every person who participates in the great live stock and meat producing industries of the United States.

CHAPTER XV

AS TO CLEANLINESS AND SANITATION

THERE has been so much said lately about the lack of care for cleanliness in packing house processes, and about the wholesomeness of the meats and food products that come through the packing houses, that I feel constrained to add something to what I have already said about the way the public health is safeguarded from the time a meat animal comes to the stock yards until it leaves the packing house as a finished product. At every step in the conversion of animals into meat the public is protected, not only by rigid government inspection of every animal before slaughter and of every carcass after slaughter, but also by the common-sense business methods of the packers themselves.

Writers and publishers who have only

the dollars-and-cents purpose to serve assume the tragic pose and unfold hair-raising tales about dark secrets of the packing industry. They make statements that should, if they were true, cause every packing house in America to be closed by law, and should convert the whole world to vegetarianism.

We are told of hidden chambers and mysterious cellars where nameless materials are worked up by secret processes into food. We are told, with a gravity that is intended to pass for sincerity, that "outsiders" are never permitted to see the inner workings of a packing house. These literary concoctions are served up with a garnishment of all the circumstantial detail that can be conceived by a dishonest mind and a feverish imagination working together. The very excess of detail, to the thinking mind, ought to be evidence of untruth.

Are these dispensers of "literature"

supernaturally gifted that they can discover in a flash the dark secrets so carefully hidden from all "outsiders" and never suspected by the thousands of packing house employees and hundreds of thousands of packing house visitors? People who would be highly indignant if one were to question their intelligence, accept these false reports because it is so much easier to believe what we are told than it is to investigate, to analyze and to think for ourselves.

If one were possessed of a mind entirely philosophical and removed utterly from personal interest in the question, he might find amusement in observing this whole performance of baiting the imaginary "beef trust" as another manifestation of the mud throwing instinct that has been characteristic of some human natures throughout the ages since time began. Unfortunately there is a serious side, a very serious side, to this performance. It does real harm, not only to the packer, but practically to the entire public.

It is a foul blow at the entire industry of raising meat animals and distributing meats and meat products, an industry that permeates every section of this country and engages, in one form or another, from farm to retail meat shop, the income producing work of a very large percentage of the population—a larger percentage probably than does any other industry in the land. It is an injustice to every man, woman or child who eats meat, because, utterly without justification, it plants in their minds a suspicion of the wholesomeness of their daily food.

The packers have nothing to hide. They hide nothing. The public knows this if it would but let itself remember facts. You who read these lines know it, if you will but forget for a moment the clamor of sensation-mongers-for-profit only. Have you ever heard of anybody being denied admission to one of the large packing houses in Chicago? Have you ever heard of anybody being restricted in packing house sight-seeing to any particular hour of the day or day of the week?

The doors of the large packing houses stand wide open to the world. There is not another industry in this country that goes farther in admitting the public to its work-rooms. Certainly no other industry, engaged in the production of food, more frankly invites visit and inspection by everybody that is minded "to see how it is done."

For more than a quarter of a century the Union Stock Yards and the packing houses have constituted the greatest show places in Chicago. It is safe to say that practically everybody coming to Chicago in that time—from other cities, from villages, from farms, from all over the world—has had it in mind to visit the stock yards and the packing houses. Most of them have

Entrance to the Union Stock Yards, Chicago, Illinois

done so. Not a day passes that we do not have them by scores; at times they come by thousands. Literally hundreds of thousands of people—I could, I believe, say millions, in truth—all meat eaters, have gone through these plants and have seen every process of converting animals into food.

Outside of the offices there is not a locked door in Chicago's Packing Town. Could the packers afford to throw their plants open to the world in this way if they had anything to conceal? Could they risk giving a bad impression of packing house processes and methods to hundreds of thousands of actual customers for their product? On the contrary, they have pursued the open house policy as good business policy. They have believed and have proved that the more the public knows about the inside of a modern packing house the less prejudice there will be against packing house food products.

But these facts, which the whole world

knows when it stops to think about them, have no restraining influence on the makers of "yellow" literature. In fact, some of them, I believe, attempt to justify their disregard for facts by holding to an academic theory to the effect that there is nothing in common between facts and truth; that truth, in other words, consists not in the literal fact of a statement, but in what the people are ready to believe and will believe.

Unfortunately a good many people will always believe anything that is persistently told them, particularly if it be about a corporation.

The business of slaughtering meat animals and converting them into food and food products is not a parlor business at its best. Yet we hear every day, from the thousands of people who visit the packing houses as sightseers, expressions of surprise that the various processes of the business are carried on in such a cleanly manner and with so little admixture of the unappetizing.

Men who are familiar with the old-fashioned ways of slaughtering and handling animals, or men who remember the stock yards and packing houses of ten or fifteen years ago, would be amazed to see what has been done since then in the way of installing improved methods. The difference between the old and the new, and the rapidity with which changes from the old to the new have had to be made, is often forgotten by supersensitive persons to whom the sight of raw meat is not pleasing.

This packing industry, as we know it today, is only about thirty years old. The packing house of thirty years ago was not much more than an old-fashioned country slaughter house multiplied in size; the appliances and methods of handling were crude. The business has grown since then with marvelous rapidity. It has been a tremendous task for the packers to keep up with the demands upon them for providing sufficient buildings and equipment of the right character. But they have kept up. The packing houses of this country will compare favorably with *any* of the places where foodstuffs are handled. Even the out-of-door work that begins the moment the cattle are landed at the yards in railroad trains has received its share of attention from the side of sanitation.

The old stock yards, opened in 1865, were not much more than a series of cattle pens that could have been constructed almost any place on the prairie with posts and fence boards. For many years they continued to be inadequately drained and poorly paved. The alleys and pens were floored over with planks. There was a pretense at draining by means of box sewers, but that was all.

During the last half dozen years the facilities for handling cattle have undergone what amounts to a revolution. The yards have been almost entirely reconstructed. Alleys and pens that were floored with planks have been torn up, and the area devoted to the yarding of cattle—almost 500 acres—has been paved throughout with vitrified brick laid on a substantial foundation. The old sewerage system has given place to a new, complete, up-to-date system. The sanitary condition of the yards, consequently, is now pronounced to be practically perfect.

The reconstruction of packing houses has been going on in the same way. New buildings are taking the place of the old just as rapidly as they can be constructed. In putting up the new buildings the very best modern methods of construction and finish are adopted throughout. The use of wood or other previous material is dispensed with as far as possible. Electrical motive power has practically supplanted other forms of power, which make it possible to run the machinery of a plant from a central station, where all the labor and dust-making and smoke-making processes can be concentrated and controlled, instead of being distributed throughout many buildings.

Criticism of the large packing houses, on the score of cleanliness and sanitation, springs largely from persons who are entirely irresponsible, and who like the "yellow" periodicals, have something to gain by attracting attention to their sensationalism. To give the slightest semblance of plausibility to their utterances, they are forced to ignore the rigid government inspection that is enforced in every large packing house, and to picture the packers as men who utterly lack ordinary business sense.

I feel safe in asserting that meats and food products, generally speaking, are handled as carefully and circumspectly in the large packing houses as they are in the average home kitchen. Put it on the basis

Government Microscopical Inspection, Union Stock Yards, Chicago, Illinois

of selfishness, if you please, and deny the packers a worthy motive, and the statement I have made is still sound.

It is good business for the packer to be careful. There is a positive dollar-and-cents benefit to him in being able to assert that his output is above reproach. He could not afford to assert this if it were not true. He would be soon found out, and the damage he would suffer from exposure would overwhelm any little momentary profit he might have reaped as the fruit of his misrepresentations.

Armour & Co., and I think most large packers, maintain a force of scientific men who are engaged in original research and experiment. An important part of the duty of these experts is to determine by frequent test that every process employed in producing articles of food shall, first of all, safeguard the wholesomeness of the article so produced.

Every detail in the handling of meats

that go into packing house food products is closely watched by trained superintendents and foremen, who are under strict instructions to prevent the slightest carelessness or the doing of anything to which exception might be taken. Scrupulous cleanliness is ordered and enforced. If this were not so, we would hardly throw open the doors of our plants, as we are doing, to visitors who have a curiosity to know what goes on inside a packing house. The tendency of packing house management is all in the direction of securing as rapidly as possible the best possible conditions surrounding the workers and the materials they handle.

It is rather late in the day, if you but think for a moment, to commence believing that canned meats and other packing house products are unwholesome. The West, the Northwest, Alaska, even the uttermost parts of the earth, have been explored and opened to civilization on Chicago canned meats. The western ranch and mining camp could not exist without Chicago canned meats. The basis of every expedition to find the north pole is Chicago canned meats. For a quarter of a century naval and military expeditions have been organized on Chicago canned meats. The wars of the world—in the Soudan, in South Africa, in China and in Manchuria—have been fought on Chicago canned meats.

The self-appointed saviors of the universe (at so much a line) sometimes point to German restrictions on meat imports from America as evidence in support of their assertions against packing house methods. This is another of their half truths. German restrictions on American meats are purely protective in deference to the agrarian influence in that nation. It has happened, too, that when German laws were barring American meats from German ports, the commissary department of the German Navy was buying American canned meats to provision his Imperial Majesty's battleships.

The question of wholesomeness and healthfulness is only part of the broader question—"quality." The large packers have grown and prospered beyond their smaller competitors not only because they have handled the business better and more economically, but because their methods of handling, and the stimulation given to the cattle industry by their facilities, has enabled the public to get a better quality of meat at the same, or a less, price than was previously paid. This has been one of the results, as stated elsewhere, of the coming of the packers, which changed raising and marketing of cattle from an adventure to a business and made cattle raising for quality profitable.

Packers are not claiming any particular virtue in this. They have done what they have done to serve their own purposes. But certainly the doing of it has not been a vice.

It has been part of the evolution of the business of the country. Just as in the transportation business the big railroads furnish better transportation facilities than the little jerk-water local line, so the large packers are able to furnish the public better meat, and furnish it regularly.

This whole question as to better prices to the cattleman, and better quality to the public, has been ably discussed in an article on "Butchers Then and Now," published in the January 1st, 1906, business review number of the *Daily Live Stock Reporter*, of Fort Worth, Texas, from which the following extracts are taken. I quote this, because live stock papers are published for, and in the interest of, cattle sellers instead of cattle buyers like the packers:

"Up to fifteen years ago the only grading of cattle was by three—a cow, a steer, and a calf; and they were in two classes—poor and fat. They were almost without exception the 'scrub or long-horns.' It is an exception to-day to find a long-horn. They sold then at from $3.00 to $8.00 per head, depending on who wanted a few head, and for what purpose. They were bought in job lots and by herds. Now they sell for from $9.00 to $60.00, single and by car-loads, at an established market whose prices are as stable as any other market, giving the producer or trader a fixed basis on which to operate.

"The NOW steer (in Texas) is fattened mainly on cottonseed hulls and meal, commodities unknown and thrown away until the packing house industry created a demand for bigger, fatter and finer meat. The NOW steer, under these conditions, weighs one thousand to one thousand two hundred pounds, while a little grasser THEN steer weighed six hundred and fifty to nine hundred pounds; and there is a large difference in the price per pound of the NOW steer, though no more time is required to produce him, but some more expense for seed and